Alternative Farming Systems in Dry Temperate Zone of Himachal Pradesh

A Case Study of Kinnaur District

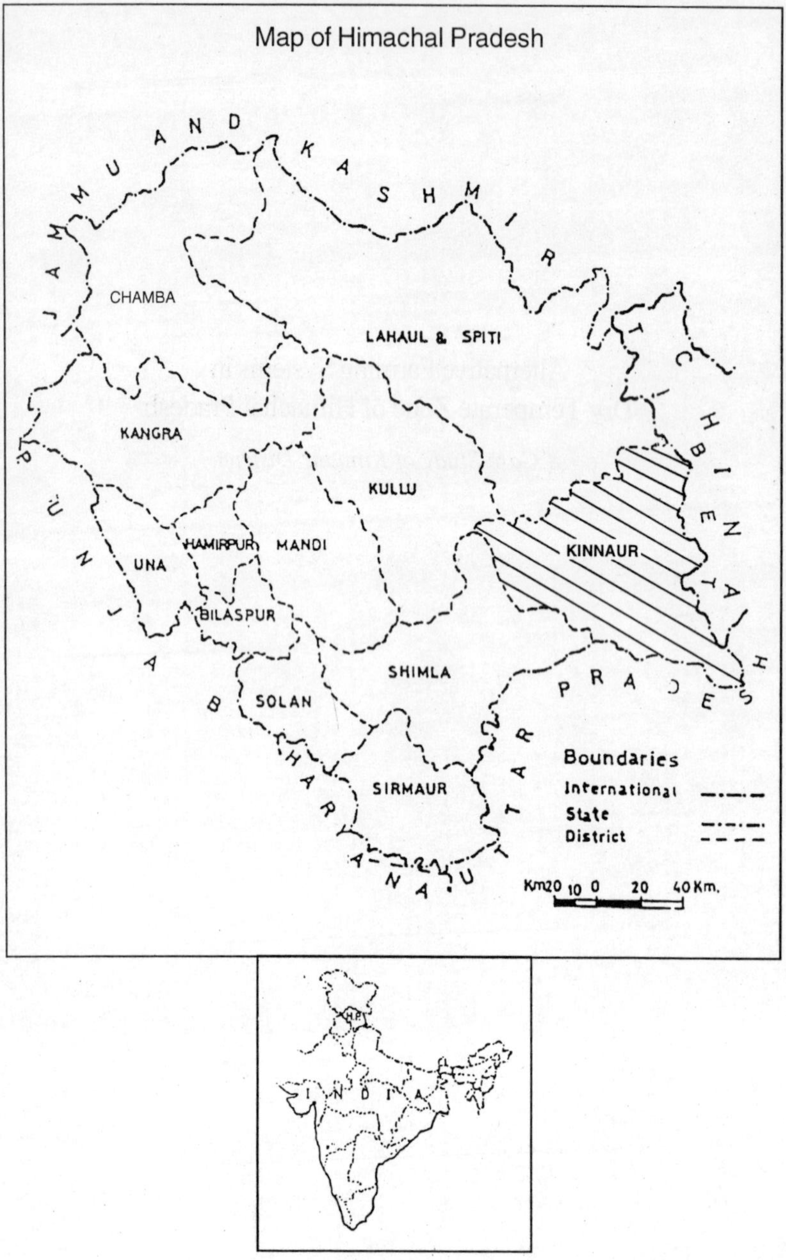
Map of Himachal Pradesh
JAMMU AND KASHMIR
CHAMBA
LAHAUL & SPITI
KANGRA
KULLU
KINNAUR
HAMIRPUR
MANDI
UNA
BILASPUR
SHIMLA
SOLAN
SIRMAUR
PUNJAB
HARYANA
UTTAR PRADESH
TIBET
CHINA
Boundaries
International
State
District
Km20 10 0 20 40Km.
H.P.
INDIA

Alternative Farming Systems in Dry Temperate Zone of Himachal Pradesh

A Case Study of Kinnaur District

A.S. RAWAT

INDUS
PUBLISHING COMPANY

ISBN 81-7387-133-7

Published by M.L. Gidwani, Indus Publishing Company
FS-5, Tagore Garden, New Delhi 110 027, and printed at
Elegant Printers, Maya Puri Indl. Area, New Delhi

Preface

The tribal regions (districts of Kinnaur and Lahaul & Spiti and Pangi and Bharmaur tehsils of Chamba district) in Himachal Pradesh, perhaps are unique and most adventurous in the world. The tribal areas of Himachal Pradesh constitute the area of 23,655 sq. km (which is 42.5 percent of the total area of H.P.) having boundaries in the north-east with China and in the north with Jammu & Kashmir. The geographical isolation and aboriginal climate make the living hard but full of adventure as the entire tribal belt is mountainous with altitude ranging from 2100 to 6000 metres above mean sea level.

India is having the largest tribal population next to Africa, constituting more than 40 million people, forming 7 percent of the country's population. Tribal population is scattered more or less in every part of the world except Europe. According to 1991 Census, 1,51,433 persons live in tribal areas in Himachal Pradesh which is nearly 3 percent of the population of the entire state.

Gaddi, Gujjar, Jad, Lamba, Khampa, Bhot, Lahaula, Pangwala, Kinnaura and Swangla are the Scheduled Tribes of Himachal Pradesh. In Himachal Pradesh, the tribal areas are the most secluded, remotest, inaccessible cold deserts with dreaded hill tracts due to mighty high mountain passes which remain covered under snow for six to seven months from Nov.-Dec. to May-June every year. Due to these reasons, tribal areas have remained neglected and lagged behind in development.

This book highlights the main problems of farmers of Kinnaur district, which is situated at longitude $77^{0}\,45'\,00''$ to $79^{0}\,00'\,35''$

east and latitude 31°05′ 35″ to 32°05′ 15″ north. It has an area of 6401 sq. km with a population of 71,270 persons.

Due to inaccessible and tough area, the research scholars are reluctant to go to these areas. Inspite of the hardships, I have tried to study at primary level the problems and constraints of the farmers of Kinnaur district. It was firmly resolved after Independence to incorporate special provisions for protection and development of the tribals. The directive contained in Article 46 of the Constitution of India clearly provides: "The state shall promote with special care the educational and economic interests of the weaker sections of the people and in particular of the Scheduled Castes and Scheduled Tribes and shall protect them from social injustice and all forms of exploitation." Remarkable efforts have been made in this direction for the tribal development during the plan periods. The plans and programmes formulated for their development can be found in the Annual Reports of the Commissioner for Scheduled Castes and Scheduled Tribes from 1950 onwards, specially the fifth five year plan marked the turning point in the history of the Scheduled Tribes. This process was further accelerated and strengthened with the introduction of the concept of Integrated Tribal Sub-Plan (ITDP). Agricultural sector has been accorded high priority in terms of fund allocation in every tribal sub-plan. It is an obvious necessity because large proportion of tribal population is dependent upon agriculture and allied activities as a source of their livelihood.

In this study it was found that the farmers of Kinnaur district are not much bothered about the input-output ratio in the farming system. This book gives some good suggestions to the farmers of this area, so that their socio-economic conditions can be changed. The secondary data were supplemented with primary data, which were collected from Department of Agriculture, Department of Horticulture, Directorate of Land Records, Directorate of Economics & Statistics, Forest Department and various published & unpublished documents.

It gives me utmost pleasure to express my gratitude and

indebtedness to Dr. Shyam Prashad, Reader in Himachal Pradesh University. No expression of thanks will be sufficient without recognition of intelligence and professional dexterity of Dr. J.P. Bhati (Professor), Dr. Dalip Thakur (Professor and Chairman) of H.P. University. I owe a deep sense of gratitude to Dr. Y.S. Negi (Reader), Dr. N.S. Chauhan (Professor) of University of Horticulture & Forestry, Nauni, Solan. I wish to record my sincere thanks to Dr. Ranveer (Research Officer) of Agro Centre, H.P. University. I extend my lively sentiments to Mr. Naginder Sharma for his contribution in my work.

My sincere thanks to the Department of Education for permitting me to complete the study.

Words are not adequate to convey the depth of my feelings of gratitude to my esteemed parents, brothers, my wife and other family members for their deep affection, inspiration, sacrifices and constant support without which this book would have merely been a dream for me.

I solely claim all the responsibility for the errors and omissions in this book.

A.S. Rawat

Contents

List of Tables

CHAPTER ONE

Introduction

Farming is composed of all the land used systems and practices where crops, livestock and other land-based activities are taken up under one management unit. The concept of farming system is quite new and is in the developmental stage. The farming system approach to agricultural development is holistic, farmer-based, inter-disciplinary, continuous, dynamic and interactive. The main factors influencing the farming system are climate, operational factors, level of socio-economic development and access to markets (Duckham and Masefield 1970, Ruthenberg 1976). Therefore, the conceptualisation of the farming system requires, an in-depth examination of all components of the system, farm technologies, management practices and socio-economic & physical environment. Of late, Indian planners have begun to realize the importance of farming system approach for the overall development of farming units, thereby raising the productivity of the whole system alongwith natural resource conservation (Planning Commission 1989, 1990).

The sustainability of agricultural production is of crucial importance for present and future generations. The ultimate goal is to develop farming system that is productive and profitable, to conserve the natural resource base and protect the environment in the long run. Thus, a location-specific understanding of the farming systems and the complex inter-relationships of natural resources with economic development is of paramount importance. Such an understanding would help policy-makers and planners to make optimal decisions with respect to decentralised and ecologically-friendly economic development in the hills.

Agriculture is the corner stone of Indian economy and engages nearly 70 percent of the work force. The pace of agricultural development was set in motion after independence through various development programmes. Of these, IADP (Intensive Agricultural District Programme) has been a big pace-setter in this regard. This programme laid the foundation of the green revolution in India which enabled the country to achieve self-sufficiency in foodgrain production.

In Himachal Pradesh, the tribal areas have been most remote and inaccessible; conditions of cold deserts and mountain topography have ensured that the tribal population and tribal areas have remained neglected and consequently have lagged behind in development.

Farming in these areas is dominated by small and marginal farmers, accounting for about 80 percent of the total cultivated land holding. In this context the application of farming system research (F.S.R.) has a great scope in the area. Further, majority of these households carry varied activities on their farms.

Historically, agricultural research and development programmes in India have concentrated on single enterprise or a single farm production system, that too in isolation. The traditional approach to agricultural development has been activity-specific. The implicit assumption is that each activity is independent of other activities and that development of individual activities would lead to overall development of the system. This ignores the notion of "wholeness" and the inter-relationships among different activities. Most of the mountain activities are linked with the plains and there is a high degree of inter-dependency among different sub-sectors of mountain farming systems (Bhati 1981, Mahat 1987, Sharma, et al. 1989). Yet, in most of the development plans regarding western Himalayan region of India, the integration of different sub-sectors of the economy is missing (Bhati 1990, Bhati and

Swarup 1985). Researchers and planners have also failed to take into account the role of mountain specificities and thus have not been able to evolve relevant and effective approaches for sustainable mountain development (Jodha 1989a, 1989b, 1990). For the hill areas, having fragile environment, system approach which takes into account the location factors and places the need of the farmer or those of his farming system at the centre of agricultural development, is quite relevant.

The current development process in the Himalayas is not only unsustainable but also incompatible with environmental conservation, because the resources essential for human development and survival are being depleted and destroyed at an ever increasing pace. There are visible indications of this process of depletion. Diversity of mountain agriculture is reduced, time spent by villagers for collection of fodder and fuel from neighbouring common property resources is longer and poverty, unemployment and outmigration of hill people are higher (Ives and Missal 1989a). The dependence of farming system on natural resources provides insight into the problems and prospects of sustainable development. The small size of land holding and meagre resource position of farmers, their heavy dependence on forestry resources is putting stress on forest wealth which will have detrimental effect on long term economic prosperity of the people of Himachal Pradesh. The nature and degree of dependence also varies according to farm activities and income status of farmers in different agro-climatic conditions (Bhati and Sharma 1996). Excess population pressure on natural resources, overstocking of pastures, neglect of development of common property resources, increased demand of wood for fuel and for packing cases of fruits and vegetables etc. are some of the main problems in the western Himalayan region of India (Bhati 1981, 1983, Bhati and Zingel 1987). Hence, there is an urgent need to evolve a strategy for sustainable agricultural development and for conserving/restoring the ecological balance in the hill areas, specially in the Himalayas.

Objectives of the Study

In order to analyse the economics of various farming systems, contribution of common property resources (CPRs) and linkages between different components of farming systems, the present study has been undertaken with the following specific objectives:

i) to identify the alternative systems of farming in dry temperate zone of H.P.;
ii) to study existing farming systems and their interlinkage in Kinnaur district of Himachal Pradesh;
iii) to analyse the dependence of farm enterprises of various farming systems on the local common property resources (CPRs); and
iv) to analyse the policy imperatives of existing farming system in the study area.

Hypotheses

In compliance with the objectives of the study, the following are the hypotheses for empirical testing:

i) that the farmers are guided by survival considerations while switching over to high value cash crops;
ii) that there are large numbers of livelihood options like crop husbandry, livestock husbandry, forestry etc. to meet their basic needs and requirements;
iii) that various farming systems are interlinked and interdependent;
iv) that common property resources being an important part of the economy of tribal region, is directly or indirectly related to all the farming sub-systems; and
v) that there are improvement in human resources, reduction in family size, changes in number and composition of animals, occupational diversification, fast growing horticultural and off-season vegetable production activities and increasing substitution of natural resources with synthetic resources.

The study has been conducted in district Kinnaur which is located in the dry temperate zone of Himachal Pradesh.

An Overview of Himachal Pradesh

The information is based on published data from various sources. It includes general background of different agro-climatic zones, land use and cropping pattern, production and crop diversification indices of Himachal Pradesh.

Geographical Setting

The state of Himachal Pradesh, nestling in the lap of the Himalayas, was born on April 15, 1948. It is situated between latitude 32° 22′ 40′′N to 33° 12′ 40′′N and longitude 75° 47′ 55′′E to 79° 04′ 22′′E with an elevation varying between 350 and 6975 metres from mean sea level. It has 12 districts, 3 divisions, 48 sub-divisions, 102 tehsils and sub-tehsils and 16,997 villages and a score of towns. It embraces an area of 55,673 square kilometres. The mainstay of the people in the state is agriculture on which 66.71 percent of its population depends. The topography being mostly hilly, the type of cultivation is terraced.

According to 1991 Census, the total population of the state is 51,70,877 with a density of 93 persons. Male female ratio works out to be 1.02 : 1. Rural population constitute 91.28 percent of total population. Scheduled Castes and Scheduled Tribes comprise 25.34 percent and 4.22 percent of total state population respectively. Out of the total population main workers account for 80.34 percent of which 50.82 percent are cultivators, 2.65 percent agricultural labourers, 25.73 percent are engaged in other activities. Other marginal workers comprise of 19.66 percent of total population and rest are non-workers. The literate population is 63.86 percent.

Operational Holdings

The total area under operational holdings according to 1990-91

Census is 8,44,213 hectares. This constitutes 15.16 percent of total geographical area. The average size of holding in the state is 1.20 hectares. There is a preponderance of small land holdings in the state. The holdings of less than one hectare accounts for 63.8 percent of total holdings which comprises 21.5 percent of total area. If small farmers are taken into account with marginal farmers then the percentage of holdings comes to 83.7 percent whereas the coverage up to 2.0 hectares size class is only 44 percent. Semi-medium, medium and large farmers covered an area of 25.7, 20.4 and 9.9 percent of total operational area in the state respectively. The corresponding figures for percentage of holdings comes to 11.4, 4.3 and 0.6 percent. This trend reveals skewed distribution of holdings in the state. With the limited land available for cultivation and relatively large population of both human and cattle, there is a heavy pressure on land. Further, only about 17.08 percent of the area is irrigated and the remaining area is subject to rainfed conditions (Directorate of Land Records).

Climate & Soil

The state experiences large climatic variations. The sharp climatic division is marked along the Pir Panjal separating the monsoon regime to its south from the non-monsoon regime to its north. In the northern latitudes, the climate is cold dry and arid varying on the Tibetan type. In the southern latitudes, it ranges from humid sub tropical to various shades of a temperate climate diversity. It brings out the contrast between the climate patterns of valleys, bottoms and the mountain tops. The state has relatively hot summers and mild winters in the south, light summers and cold winters in the interior and severe winters and pleasant summers in inner parts in the north. In the inner zone, summers are experienced late in July-August and temperature drops rapidly after October. Lahaul in this part has an average annual temperature of 10.8°C and a diurnal range of 1 to -4°C even in the month of October. In Spiti, a rise or fall either way up to 23.4°C within 12 hours has been recorded. Small glacial streams in these areas freeze near their sources not only in winters but also during the night in other seasons. Taking both temperature and rainfall into consideration the climatic typology of the state is as given in Table 1.1.

Table 1.1. Climatic typology of the state

Climatic group	Districts
Wet sub-tropical	Kangra, Mandi, Bilaspur and lower Sirmaur.
Cold humid temperate	Inner and higher elevations of outer Himalayas —Chamba, Kullu, Shimla and Solan.
Cold moist temperate	Tissa, Bhangal, Bharmaur, Pangi, Rampur, Kalpa tracts and some parts of Mandi.
Cold dry temperate	Lahaul & Spiti and Kinnaur.

Source: Gangopadhyay, 1991.

The state has the highest and lowest average annual rainfall of about 300 cm at Dharamshala and 35 cm at Pooh, which is further reduced to 18 cm in the interior of Spiti. For the state as a whole, the average annual rainfall is 150 cm. While large proportion of it is concentrated in summer season in the outer zone, the monsoon comes only just in the form of a misty drizzle in parts of the inner zone, which mostly gets moisture in winter in the form of snow.

A wide variation exists with regard to soil structure, texture and fertility as well. The soil of the state can be broadly classified into eight main groups. Loam with varying degrees of stone content, clay loam with lesser proportion of stones, heavy soils with large clay contents, reddish to red soil with stones in varying proportions, slightly black soils with smaller proportion of stones, very stony soil, greyish and slightly chalky soil with an admixture of humus. The soils in most of the areas are near-neutral and acidic in the high rainfall regions. On the basis of agro-climatic conditions, the state of Himachal Pradesh is divided into four zones.

Table 1.2. Agro-climatic zones of Himachal Pradesh

Name of the zone	Areas under the zone
1. Sub-montane and low hills sub-tropical zone (zone I)	Una, Hamirpur, Bilaspur district and parts of Solan, Sirmaur, Kangra and Chamba district.
2. Mid-hills sub-humid zone (zone II)	Palampur, Kangra, Rampur tehsil and parts of Solan, Bilaspur, Kullu, Chamba, Sirmaur and Mandi districts.
3. High hills wet temperate zone (zone III)	Parts of Shimla, Mandi, Solan, Kullu, Kangra, Sirmaur & Chamba districts.
4. High hills dry temperate zone (zone IV)	Kinnaur, Lahaul & Spiti and Pangi sub-division of Chamba district.

Salient Features of Different Zones

Planning Commission has divided Himachal Pradesh into four physiographic groups/agro-climatic zones for the purpose of regional planning. These zones have been marked on the basis of vegetation types, geographic and climatic characteristics. The information/important features pertaining to different agro-climatic zones have been given in Tables 1.2 and 1.3.

Characteristics of the Study Zone (Dry Temperate Zone)

Entire Kinnaur and Lahaul & Spiti districts, Pangi and Bharmaur tehsils and now sub-tehsil Holi of Chamba district constitute the tribal areas in the state, fulfilling the minimum criterion of 50 percent Scheduled Tribe population concentration in a community development block.

These areas have been declared as Scheduled Areas under the Fifth Schedule of the Constitution by the President of India as per the Scheduled Areas (Himachal Pradesh) Order, 1975 (Co 102) dated 21st November 1975.

Population

Himachal tribes form the major part of north-western region. The Scheduled Tribes inhabiting Himachal are the Gaddis, Gujjars,

Agro-Climatic Zones of Himachal Pradesh

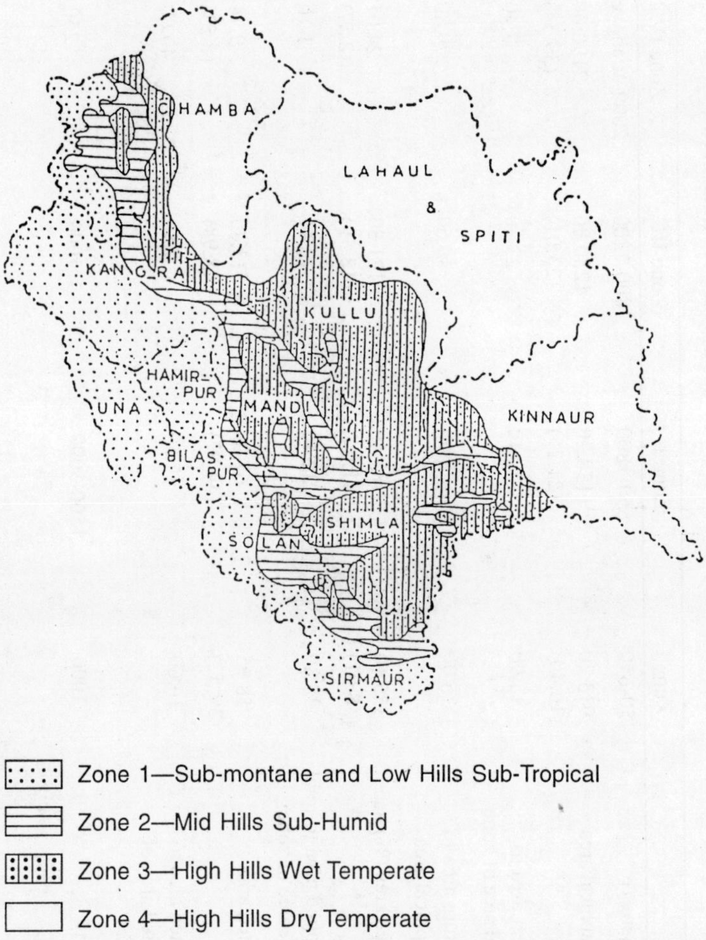

Table 1.3. General features of agro-climatic zones of Himachal Pradesh

Particulars	Zone I	Zone II	Zone III	Zone IV	H.P.
Altitude (metres)	350-650	650-1800	1800-2200	2000-& above	–
Total geographical area (%) hac.	913.20 (16.40)	1138.20 (21.25)	1280.90 (23.01)	2190.00 (39.34)	55673 (100)
Percent forest area to geographical area of zone	13.90	40.40	23.30	7.40	–
Percent pasture area to total geographical area	26.00	28.60	49.60	79.80	–
Total cropped area (%) hac.	355.10 (38.01)	383.40 (41.04)	171.80 (18.39)	24.01 (2.57)	934.30 (100)
Percent cropped area to total geographical area of zone	38.89	32.40	13.41	1.10	16.78
Gross irrigated area (000) hac.	58.95 (39.52)	67.09 (44.97)	13.40 (8.98)	9.74 (6.53)	149.18 (100.00)
Gross irrigated area to total cultivated area of zone	16.60	17.50	7.80	40.60	15.97
Rainfall (mm)	1000	1500-3000	1000	250	–

Contd.

Table 1.3 continued

Particulars	Zone I	Zone II	Zone III	Zone IV	H.P.
Livestock	Buffalo, cow, bullock, sheep & goat, donkey.	Cow, buffalo, bullock, sheep & goat.	Sheep & goat, cow, bullock, horse, mule etc.	Sheep & goat, yak cow, mule, horse, donkey.	–
Major cereal crops	Wheat, barley, pulses, mustard, potato, onion, garlic, maize, paddy.	Wheat, barley, lentil, mustard, maize, paddy, mash & bean.	Wheat, barley, lesser millets, buck-wheat & maize.	Wheat, barley, buck-wheat, minor millets (ogla/phapra).	–
Horticultural crops	Mango, citrus, ber, fig, loquat, litchi, papaya.	Peach, plum, apricot, pear, pecannut.	Apple, pear, cherry, walnut, almond, chestnut.	Apple, pear, walnut, almond, chestnut.	–
Vegetables	Tomato, bhindi, capsicum, cauli-flower, bean.	Tomato, green peas, capsicum, cauliflower, bean.	Cauliflower, cabbage, green peas, bean.	Potato, green peas, cauliflower, cabbage, chilly.	–
Main forest vegetation	Acacia, shisham, ber, peepal, bardh, tooni, beol, kharak, fig.	Ban, chir pine, beol, tooni, kharak, khair.	Deodar, kail, chir pine, oak, rhododendron.	Deodar, kail, oak, moru, tooni, birch tree, reonsh.	–

Jads, Lambas, Khampas, Bodhs or Budhs, Kinnauras or Kinners, Lahaulas, Pangwalas and Swanglas. The tribal area covers 42.49 percent of total geographical area with 2.93 percent of the total population of the state. These areas are the most sparsely populated pockets with an average density of 6 persons per sq. km as against 93 persons in the state as a whole. The entire population in these tribal areas is rural. The decennial rate of growth of population in the tribal areas between 1981 and 1991 was 13.1 percent as against 20.8 percent in the state.

Of the total population in these areas, Scheduled Tribes account for 69.3 percent, Scheduled Castes for 17.9 percent and the remaining 12.80 percent belong to other categories. Kinnaur tribals account for 55.6 percent of its total population. The number of females per thousand males are 868 in tribal areas as

Table 1.4. Demographic features of Kinnaur, tribal areas and H.P.

Particulars	Unit	Year	Kinnaur	Tribal areas	Himachal Pradesh
1. Geographical area	Sq. km	1991	6401	23655	55673
2. Population	No.	1971	49835	114264	3460334
	"	1981	59547	133847	4280818
	"	1991	71270	151433	5170877
3. Density of population	Per sq. km	1981	9	6	77
	"	1991	11	6	93
4. Sex ratio	No.	1981	885	854	973
	"	1991	856	868	976
5. Decennial population growth	%age	1981-91	19.7	13.1	20.8
6. Percentage of S.T. population	%age	1971	68.4	72.9	4.1
	"	1981	74.9	77.5	4.6
	"	1991	55.6	69.3	4.2
7. Percentage of S.C. population	%age	1971	19.4	12.2	22.2
	"	1981	10.6	8.9	24.6
	"	1991	29.9	17.9	25.3
8. Urban population	No.	1991	–	–	449196
9. No. of villages	No.	1981	77	772	18721
	"	1991	662	1423	19388

Himachal Pradesh
Tribal Areas

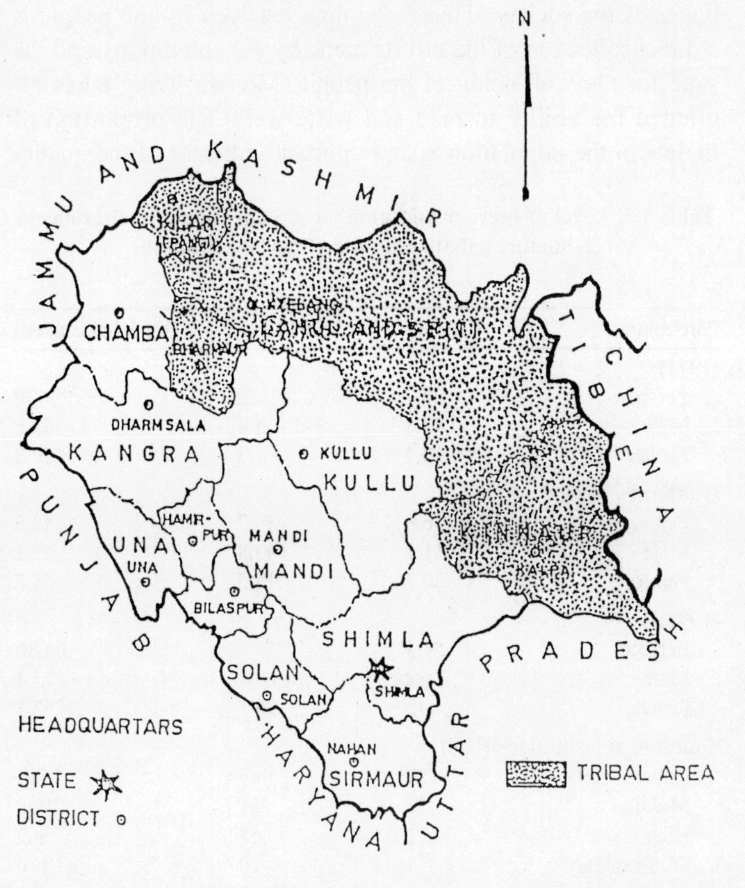

compare to 976 for the state. Similar figure for Kinnaur stands at 856 which is lower than the state as well as the tribal areas as a whole. The other demographic features of tribal areas have been presented in Table 1.4.

From the point of view of development, the quality of population is no less important than the numbers. Education ranks among the major tasks and educational progress constitutes chief hopes of the society. Thus, schooling received by the people is a direct reflection of the efforts made by the community and the state for the well-being of the people. Modern living takes for granted the ability to read and write well. The proportion of literate in the population is an important indicator of the quality

Table 1.5. Level of literacy and number of educational institutions in Kinnaur, tribal areas and Himachal Pradesh

(%)

Particulars	Kinnaur	Tribal areas	Himachal Pradesh
(1971)			
Persons	27.7	22	32
Male	43.3	34.8	43.2
Female	10.4	7.3	20.2
(1981)			
Persons	36.8	30.7	42.5
Male	51.1	43.8	53.2
Female	20.7	15.5	31.5
(1991)			
Persons	58.4	53.4	63.86
Male	72	68.1	75.4
Female	42	36.1	52.1
Educational institutions (No.)			
Primary (1993-94)	153	487	7566
Middle	27	61	1008
High	21	53	862
Sr. Secondary	4	10	170
Colleges (1996)	1	2	29

Sources: 1. Census of Himachal Pradesh.
2. Statistical Outline of Tribal Areas, 1995.

of population. In tribal areas 53.40 percent of the people were literate in 1991 which is lesser than the state literacy rate (63.9%) during the same period. In tribal areas the highest literacy was among the people of Kinnaur (58.4%). Literacy amongst females is much less than that amongst males in all the areas under study. Reasons for the situation are the same as observed in any other part of the country i.e. females are mostly kept confined to household chores and field activities and it is the privilege of the men-folk alone to have some education and if possible to pursue for a salaried job. The situation has gradually been changing in favour of females. More details of the number of educational institution etc. are presented in Table 1.5.

Distribution of Workers According to Occupation

The economy of an area depends much on the availability of active workers. This can be estimated through decennial Census data and from various rounds of National Sample Surveys. Here, the data provided in Census reports of 1991 have been utilised. According to 1991 Census most workers comprised of cultivators, agricultural labourers, those involved in household industries and other workers including factory workers, plantation workers, those in trade and commerce, business, transport, mining, construction and all Government and Municipal employees. In this Census, another category of workers i.e. marginal workers has also been introduced to provide an estimate of under-employed persons. The concept used during 1991 Census is definitely an improvement over the earlier ones. The workers who are chronically unemployed and are seeking work, are not mentioned separately in the last Census. However, there is reference to this category but it is clubbed with the category of non-workers.

Of the total tribal population of 81,155 people 83.69 percent are main workers and 16.31 percent are marginal workers. In case of Himachal Pradesh as a whole the proportion of main workers in the total workers is lesser (80.34%) than the tribal areas. In Kinnaur the main workers constitute about 90.26

percent of the total workers. The main workers who have worked for six months or more in a year are taken as fully employed, then the extent of under-employed in tribal areas and state would be around 16 percent and 20 percent of the total population respectively. In Kinnaur the situation is much better and only 9 percent of the total population is under-employed. However, in actual practice all the main workers cannot be taken as fully employed. Therefore, the extent of under-employment will definitely be higher than the estimates based on marginal workers only (Table 1.6).

Table 1.6. Distribution of workers by occupations in Kinnaur, tribal areas and H.P. during 1991

(Nos.)

	Occupation	Kinnaur	Tribal areas	H.P.
1.	Cultivators	18374 (49.18)	39385 (48.53)	1125311 (50.82)
2.	Agricultural labourers	1729 (4.63)	2728 (3.36)	58668 (2.65)
3.	Household industry, manufacturers, processing, serving & repair	944 (2.52)	1303 (1.61)	25454 (1.15)
4.	Other workers	12676 (33.93)	24508 (33.20)	569667 (25.73)
5.	Total main workers	33723 (90.26)	67916 (83.69)	1779100 (80.34)
6.	Marginal workers	3638 (9.44)	13239 (16.31)	435279 (19.66)
7.	All workers	37361 (100.00)	81155 (100.00)	2214379 (100.00)
8.	Non-workers	33909	70278	2956498
9.	Total population	71270	151433	5170877
10.	Percentage of workers in population	52.42	53.59	42.82

Land Utilization Pattern

The total geographical area of tribal region in the state is 23,654 sq. km (42.5% of total area of the state). Permanent pastures and grazing lands account for 65.46 percent of this area. These facts of inter alia the high proportion of workers depending on agricultural pursuits, suggest that it is not cultivation of crops alone on which the people of tribal areas should generally be depending. The type of livestock maintained justifies the high proportion of permanent pastures and other grazing land which are necessary to maintain the flocks of sheep and goats.

The area sown more than once in the region is also less than the state figures. Therefore, the cropping intensity in tribal areas

Table 1.7. Land utilization pattern in Kinnaur, tribal areas and H.P. during 1992-93

(000 ha)

Land utilization	Kinnaur	Tribal areas	H.P.
1. Total geographical area	640.1	2365.4	5567.3
2. Forest	22.34	282	1038
3. Land not available for cultivation	42.88	51.03	347.39
(a) Barren and unculturable land	33.43	39.24	146.08
(b) Land put to non-agri uses	9.45	11.79	201.31
4. Other uncultivable land including current fallow	172.47	348.00	1366.13
(i) Culturable waste	5.89	6.84	120.36
(ii) Permanent pasture and other grazing land	166.49	340.96	1202.97
(iii) Land put under miscellaneous tree crops not included in net area sown	0.09	.20	42.80
5. Fallow land	2.01	2.71	71.35
(i) Current fallow	0.40	1.04	48.52
(ii) Other fallow	1.61	1.67	22.83
6. Net area sown	7.49	14.69	572.56
7. Area sown more than once	1.72	3.20	400.05
8. Total cropped area	9.21	17.89	972.61
9. Cropping intensity	122.96	121.78	169.87
10. Net irrigated area	4258	8826	98826

Source: Census of India, 1991. Directorate of Land Records, Shimla.

is lower (121.78%) than the state average (169.87%). The low cropping intensity in tribal areas is due to generally prevalent inclement weather. In most parts of the region, only one crop is grown during the whole year. Forest area accounts for 26.70 percent of the total area and all the areas not necessarily covered by forests, is under the control of forest department. Other uses of land such as fallow lands and area not available for cultivation etc. is negligible (Table 1.7).

Land Holding Size and Cropping Pattern

As has already been observed that the total arable land in the tribal area is scarce. Therefore, the proportion of small and marginal holdings is high. As many as about 60 percent of the total holdings are below one hectare and cover about 24 percent of the area. The holdings with farm size upto 2 hectares are 23 percent of the holding, covering 28 percent of the total area. These figures indicate that the farmers of this area are at much disadvantage than the state in regard to land holdings. This fact, as also the large dependence of the people on agricultural pursuits, suggests that the people of the area have something else to depend on rather than land holdings alone. This obviously leads us to conclude that livestock sector must be strong to supplement the farm household incomes (Tables 1.8 and 1.9).

The soils are loamy sand to sandy loam in texture and neutral to slightly alkaline in reaction and low in fertility. Practically no crop can be raised without irrigation. Gravitational channels (kuhls) are the only source of irrigation. The soil erosion and

Table 1.8. Important crop rotations of dry temperate zone

Dry farming area	Irrigated area
Smaller millets–fallow	Barley–Buckwheat–Fallow
	Potato–Fallow
	Wheat–Fallow
	Potato–Wheat/Barley
	Rajmash–Fallow
	Peas–Fallow

Table 1.9. Different farming situations in dry temperate zone

Sr. No.	Farming situation	Crops grown
1.	Sharbo wet-dry zone	Raisin grapes, almond, walnut, apricot, apple, chilgoza, cabbage, sugarbeat, chicory, rajmash and trees of pinus, salix, alnus.
2.	Kalpa wet-dry zone	-do-
3.	Sangla complete dry zone	Buckwheat, phapra, kala zira, saffron, apple, apricot, pulses and millets.
4.	Extreme dry zone of Tabo	Millet, wheat, barley, vegetables and vegetable seed production.
5.	Leo dry zone of Tabo	Millet, wheat, barley, vegetable seed production.
6.	Kukumseri dry temperate	Buckwheat, kuth, seed potato, almonds, walnut, hazelnuts, raisin grapes, prunes, apricot, chicory, kala zira, hops, wheat, beans, pea, saffron and vegetables.

Source: Gangopadhyay, 1991.

water management are the main problems confronted by the farmers.

Cereals are the predominant crops accounting for at least 75 percent of the total cropped area in the tribal region. Comparative figures of area under cereals in the tribal areas and the state as a whole show that the area under these crops is higher in other areas as compared to tribal areas of the state. Next in importance are vegetables. This area is best suited for growing off-season vegetables. The climate and soil is most suitable for growing some of the vegetables of good quality. The vegetables grown in this area have great demand in different markets.

Fruit Crops

Agro-climatically, this region of the state is also suitable for growing horticultural crops. The proportionate area under fruits and vegetables in tribal areas is higher than that of the state as a whole. But due to the absence of needed infrastructural facilities like transportation, storage, processing etc., the potential has not been tapped. The infrastructural facilities are even more important for fruits and vegetables because these are more perishable and fragile as compared to other agricultural products. Furthermore, the proportion of marketable surplus in fruits and vegetables is much higher than other crops.

Horticulture is the fastest growing activity in hilly areas of Himachal Pradesh. It has already grown into a dominant export base of the state. However, the tribal region is still to catch up in this matter.

The fruit acreage in tribal areas is 8289 hactares (1992-93). Proportionately area under fruits is higher in tribal region than the proportionate area of fruits in the state as a whole. In tribal areas, Kinnaur district dominates in this matter with 6127 hectares under fruit crops.

Livestock

The contribution of livestock wealth to the tribal economy is quite significant. The main occupation of nomadic and semi-nomadic people of the tribal areas is pastoralism. These areas account for about 16.84 percent of the total sheep and 9.29 percent of the total goats in the state. Sheep and goat-rearing is a hereditary occupation of these tribal people. The high mountain pastures provide excellent grazing facilities during the summer months. Though rainfall in these areas is scanty, the winter snow provides sufficient moisture to the soil. Sheep in this region have superior fleece with higher percentage of fine undercoat. Due to heavy snowfall during winters, the flock masters have to take their flocks to far-off places in the lower hills. In many cases the state border is to be crossed in search of fodder.

The changes in the livestock of Himachal Pradesh during 1972 to 1992 are shown in Table 1.10. Maximum growth was noticed in the equines (horses, ponies, mules and donkeys) population, which increased at the rate of 2.43 percent per year, followed

Table 1.10. Composition of livestock (in Nos.) in Himachal Pradesh (compound growth rate)

(Nos.)

Type of livestock	1972	1977	1982	1987	1992	Annual CGR (%)
Cows	673718	685436	693197	732207	778957	0.72
Bullocks	824147	825388	853810	892205	848254	0.27
Youngstock	599840	595396	626556	620403	524445	-0.45
Sub-Total	**2997705**	**2106220**	**2173663**	**2244815**	**2151656**	**0.23**
She Buffaloes	361528	384497	413595	462392	472294	1.49
He Buffaloes	14498	12278	12811	101344	7832	1.84
Youngstock	167861	163231	190009	231255	220798	1.87
Sub-Total	**543887**	**560006**	**616415**	**794991**	**700924**	**1.79**
Sheep	1039197	1055005	1090322	1054182	1074345	0.13
Goats	906415	1035337	1059862	1120139	1115591	1.01
Sub-Total	**1945612**	**2090342**	**2150184**	**2174321**	**2189936**	**0.56**
Horses, Ponies, Mules, Donkeys	28138	29147	35962	51263	37640	2.43
Yaks	4589	3474	3495	2155	5314	-0.36
Camels	885	872	714	327	375	-4.73
Pigs	2906	5165	8107	17976	7095	7.08
Sub-Total	**36518**	**38658**	**48278**	**71721**	**50424**	**2.69**
Total livestock	**4623722**	**4795226**	**4988540**	**5285848**	**5092940**	**0.59**

Source: Livestock Censuses, Directorate of Land Records, Himachal Pradesh, Shimla.

by buffaloes whose population increased by 1.79 percent per annum. Cattle population of 1945 thousands in 1972 rose to 2189 thousands by the year 1992, showing a growth rate of 0.56 percent per annum. Respective growth rates of sheep and goats population were 0.13 percent and 1.01 percent per annum during the 20-years period. During 1992 Census, sheep and goats population stood at 10.74 lakhs and 11.15 lakhs respectively.

Composition of different livestock in Himachal Pradesh and

Table 1.11. Composition of livestock (in %) in Himachal Pradesh

(%)

Type of livestock	1972	1977	1982	1987	1992
Cows	14.58	14.29	13.90	13.85	15.29
Bullocks	17.82	17.21	17.12	16.88	16.66
Youngstock	12.97	12.42	12.56	11.74	10.30
Sub-Total	**45.37**	**43.92**	**43.58**	**42.47**	**42.25**
She Buffaloes	7.82	8.02	8.29	8.75	9.27
He Buffaloes	0.31	0.26	0.26	1.92	0.15
Youngstock	3.63	3.40	3.81	4.48	4.48
Sub-Total	**11.76**	**11.68**	**12.36**	**15.05**	**13.76**
Sheep	22.48	22.00	21.86	19.94	21.09
Goats	19.60	21.59	21.25	21.19	21.91
Sub-Total	**42.08**	**43.59**	**43.11**	**41.13**	**43.00**
Horses, Ponies Mules, Donkeys	0.61	0.61	0.71	0.97	0.74
Yaks	0.10	0.07	0.07	0.04	0.10
Camels	0.02	0.02	0.01	(Neg.)	0.01
Pigs	0.06	0.11	0.16	0.34	0.14
Sub-Total	**0.79**	**0.81**	**0.95**	**1.35**	**0.99**
Total livestock	**(100.00)**	**(100.00)**	**(100.00)**	**(100.00)**	**(100.00)**

Source: Livestock Censuses, Directorate of Land Records, Himachal Pradesh, Shimla.

tribal areas of the state for the years 1972, 1977, 1982 and 1992 indicates that the total population of livestock increased by 7.97 percent and 2.12 percent during 1972 to 1982 in the state and tribal areas respectively. In the total livestock population of Himachal Pradesh, the proportion of cattle is highest, i.e. about 43 percent, followed by sheep, and goats which accounted for 21 and 22 percent respectively during 1992. In the tribal region of the state, the proportion of sheep is highest i.e. 43 percent during the 1992 Census. The cattle account for 67 percent of the total livestock population in the tribal areas during the same year. Though there is an increase in the absolute number of cattle

Table 1.12. Composition of livestock (Nos.) in tribal region of Himachal Pradesh (compound growth rate)

(Nos.)

Type of livestock	1972	1977	1982	1987	1992	Annual CGR (%)
Cows	79652	81682	83059	70573	83695	-0.09
Bullocks	66913	68508	72742	86215	76377	0.01
Youngstock	60224	60497	63654	59431	55280	-0.37
Sub-Total	**206789**	**210687**	**219455**	**216219**	**215352**	**0.21**
She Buffaloes	15500	15624	16795	15153	17072	0.33
He Buffaloes	424	413	379	1268	436	2.50
Youngstock	5260	4467	5560	6270	6078	1.30
Sub-Total	**21184**	**20504**	**22734**	**22691**	**23586**	**0.64**
Sheep	320120	342650	329695	326760	329293	0.02
Goats	160292	160071	168848	178419	177500	0.63
Sub-Total	**480412**	**502721**	**498543**	**505175**	**506793**	**0.22**
Horses/Ponies Mules /Donkeys	6710	6500	8083	13703	9768	3.23
Yaks	4589	3474	3495	2155	5301	–0.37
Camels	–	–	–	–	–	–
Pigs	3	72	122	438	140	31.67
Sub-Total	**11302**	**10046**	**11700**	**16296**	**15209**	**2.27**
Total livestock	**719687**	**743958**	**752432**	**760385**	**760940**	**0.27**

Source: Livestock Censuses, Directorate of Land Records, Himachal Pradesh, Shimla.

in the state, the proportion has declined in total livestock population during 1972 to 1992 period. In tribal areas the number as well as proportion of cattle increased during the same period.

The State Government has launched many programmes to improve the quality of sheep stock. The proportion of sheep in total livestock in tribal areas also decreased marginally from 53.32 percent to 52.59 percent during the same period.

Table 1.13. Composition of livestock in tribal region of Himachal Pradesh (Percent)

(%)

Type of livestock	1972	1977	1982	1987	1992
Cows	11.07	10.98	11.04	9.28	11.00
Bullocks	9.30	9.21	9.67	11.34	10.04
Youngstock	8.37	8.13	8.45	7.82	7.26
Sub-Total	**28.74**	**28.32**	**29.16**	**28.44**	**28.30**
She Buffaloes	2.15	2.10	2.26	1.99	2.24
He Buffaloes	0.06	0.06	0.05	0.17	0.06
Youngstock	0.73	0.60	0.74	0.82	0.80
Sub-Total	**2.94**	**2.76**	**3.05**	**2.98**	**3.10**
Sheep	44.48	46.06	43.80	42.97	43.27
Goats	22.27	21.52	22.43	23.46	23.33
Sub-Total	**66.75**	**67.58**	**66.23**	**66.43**	**66.60**
Horses, Ponies Mules, Donkeys	0.93	0.87	1.08	1.81	1.28
Yaks	0.64	0.47	0.46	0.28	0.70
Camels	–	–	–	–	–
Pigs	(Neg.)	(Neg.)	0.02	0.06	0.02
Sub-Total	**1.57**	**1.34**	**1.56**	**2.15**	**2.00**
Total livestock	**(100.00)**	**(100.00)**	**(100.00)**	**(100.00)**	**(100.00)**

Source: Livestock Censuses, Directorate of Land Records, Himachal Pradesh, Shimla.

Goat population, which is discouraged by the government because they cause maximum damage to vegetative cover, surprisingly registered an increase in absolute number as well as in proportion in the total livestock population during the period of one decade. This may be due to the increasing demand of goat meat by the people in the state. Goats when let into forests cause maximum damage as they are voracious browsers and are very fond of nibbling young seedlings and growing buds.

The population as well as proportion of equines in the state and tribal area increased during 1972 to 1992. At present, considerable number of people in the hills particularly in remote areas are depending on horses, ponies, mules and donkeys in various degree, as a part of their means of sustenance. The donkey in particular is an important animal in the economy of landless labourers, small and marginal farmers and other poor sections of the community in this tribal area. Inspite of a marked increase in the roads and transport facilities, the number as well as proportion of equines has increased. This marked increase is due to the increase in production of commercial crops, e.g. apples and potatoes which have come up in a very big way and are to be carried from far flung remote areas to roadheads, generally by non-wheeled means for onward transportation to various markets.

Yak is the most important animal for the people living on higher altitude particularly in tribal region of the state. It is the only species of animal which provides milk, meat, hide, wool and also used for transport at higher altitudes where no other ruminant can thrive economically. Its ability to withstand low temperatures, snowy environment and survivability on coarse fodder are unique characteristics. The yak can eke out its living out of stubs and can dig out roots of vegetation even buried under snow as deep as 10-15 cms. It is a high climber and can plough through snow in winter. They thrive well at an altitude around 10,000 feet above msl, but in summer they may go in search of food beyond 17,000 feet. These qualities of yak have functional comparison with the camels and can be safely termed as ship of the highland or camel of the cold desert. It provides draught power to meet the requirements of agricultural operations, serves as a beast of burden and also for riding at times. Yak meat is considered to be quite delicious. The yak hair is generally used to make mattresses, rugs, ropes, tents and the bushy tail is used as 'chaunri' in religious places. Yaks are also reared for cross-breeding with cows. The population of yaks in the state increased from 4589 during 1972 to 5303 in 1992. Nearly 82 percent of the total yak population of the state is found in tribal areas of the state.

Areas of tribal region constitute about 72 percent of the total livestock of the region. Of the total population of sheep in tribal region nearly 68 percent are reared in Kinnaur area. Nearly 83 percent of the total goats population is found in Kinnaur area of the tribal region. This suggests that Kinnaur area is most important part of Himachal for rearing sheep and goats.

From the above it may be concluded that the potential of livestock farming in tribal areas is very high. Unlike the cattle and buffaloes, the number of sheep and goats per person in Kinnaur is many times more than that in tribal region and the state as a whole. This is the fact that the quality of sheep and goats in Kinnaur is better than that of other livestock in the area and suggests that the dependence and emphasis of the people is relatively more on these animals.

General Features of the Study Area (Kinnaur District)

Location

District Kinnaur is situated at longitude 77° 45′ to 79° 00′ 35″ east and latitude 31° 05′ 35″ to 32° 05′ 15″ north. It has an area of 6401 sq. km with a population of 71,270 persons. The consequent population density is only 11 persons per sq. km. This is entirely a tribal district and has a dominant population of Scheduled Caste and Scheduled Tribes. Of the entire population 30 percent belong to Scheduled Castes and 55.60 percent belong to Scheduled Tribes. In the district there were 33,723 workers and 3638 marginal workers. According to 1991 population Census there were 18,374 farmers in district Kinnaur and 1729 agricultural labourers (Table 1.6). Agriculture and horticulture are the main source of livelihood in this district. Its southern boundary touches Uttarkashi district of Uttaranchal and Chirgaon tehsil of Shimla district.

Climate

The climate resemble arid Tibetan type, characterised by absence

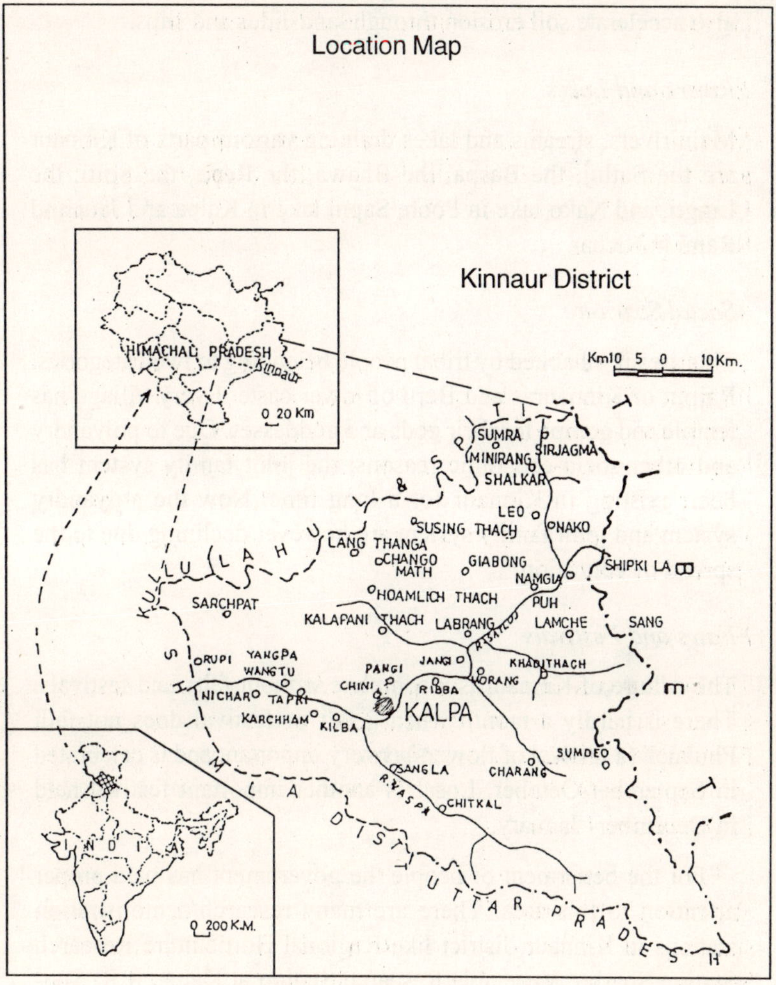
Location Map
Kinnaur District
HIMACHAL PRADESH
Kinnaur
0 20 Km
I N D I A
0 200 K.M.
Km10 5 0 10Km.
SUMRA
SIRJAGMA
MINIRANG
SHALKAR
LEO
NAKO
SUSING THACH
LANG THANGA
CHANGO MATH
GIABONG
SHIPKI LA
NAMGIA
PUH
HOAMLICH THACH
SARCHPAT
KALAPANI THACH
LABRANG
LAMCHE
SANG
R. SATLUJ
JANGI
KHADITHACH
RUPI
YANGPA
WANGTU
PANGI
CHINI
RIBBA
MORANG
NICHAR
TAPRI
KALPA
KARCHHAM
KILBA
SUMDEO
SANGLA
CHARANG
R. VASPA
CHITKAL
KULLU
LAHUL & SPITI
SHIMLA DIST
UTTAR PRADESH

of monsoon rains during summer. Snowfall is heavier in the whole Kinnaur-Satluj catchment during December to March. The rapid melting of snow causes frequent avalanches and glaciers which cause severe wind storm damaging and uprooting forests. They also accelerate soil erosion through landslides and slips.

Rivers and Lakes

Main rivers, streams and lakes draining various parts of Kinnaur are the Satluj, the Baspa, the Bhawa, the Ropa, the Spiti, the Lingti, and Nako lake in Pooh, Sapni lake in Kalpa and Jani and Rami in Nichar.

Social Structure

This area is inhabited by tribal people belonging to two categories, Rajput or Kinnauras and Beru of lower caste. Every village has temple and gompa for their gods and goddesses. Due to polyandry and other socio-economic reasons, the joint family system has been existing in Kinnaur for a long time. Now the polyandry system and joint family system are however declining due to the spread of education.

Fairs and Festivals

The culture of Kinnaur is incomplete without fairs and festivals. There is hardly a month when a fair or festival does not fall. Phulaich (a festival of flowers) is very important and is celebrated in September-October. Losar is another important festival held in December- January.

For the betterment of people the government has paid proper attention to this area. There are many research/demonstration stations in Kinnaur district like Regional Horticulture Research Station Sharbo, Vegetable Research Station at Kalpa, R.R. Sub-Station Tabo (Spiti), Krishi Vigyan Kendra Peo, P.C.D.O. (Progency-cum-Demonstration Orchard) Boktu, Research Station Leo and Research Station at Sangla for Zira and Kesar cultivation.

CHAPTER TWO

Studies on Farming Systems and CPRs

In this chapter an attempt has been made to summarise the results of the studies undertaken by various researchers within and outside the country on various aspects of farming system and role of CPRs and its management. The studies reviewed are classified into three categories: (i) those dealing with linkages of farming systems, (ii) those which highlight the role played by common property resources in farming systems, and (iii) those which relates to strategies for common property resource management.

Characteristics and Linkages of Farming Systems

Parasnis (1976) studied linkages between the forestry and agriculture by using input-output model. He found the potential of growing food crops and forestry crops, provided there is adequate integration of these enterprises. Singh (1992) conducted the study in Kangra district of Himachal Pradesh. He revealed that there is a strong bondage of input supply among the different farm sectors.

Batini (1977) analysed the extent of integration, resource allocation and optimisation in the farming system in general land use scheme. How should these two activities be integrated; in what proportion; which areas favour each and how is best mix obtained. These are important considerations not only for social and economic values but also for other factors, such as to aesthetic quality of landscape. The author emphasized that apart from technical the large scale development of integrated agriculture and forest farming has considerable social and economic implications too.

Learmonth & Robbette (1977) examined the comparative economics of integrated enterprises with single enterprise in the Ballarat district of Victoria (Australia). The important results that emerge from the study are:

- that the integrated enterprise is superior to the sheep enterprise under all physical and financial assumptions and discount rates;
- the 30 year rotation of integrated enterprise is much more profitable than all other options at all levels of price and discount rates.

Newbold (1977) found that some form of integrated land use combining forestry and agriculture must be devised for U.K. if conflicts are not to arise and if agriculture is not to decline. He stressed that the farmer and estate owner need more information and guidance plus the continued provision of suitable grant and tax schemes to encourage them to regard tree as a crop and to integrate their use on or around the farm. More research is required into the possible benefits of dual land use over a range of British conditions.

Gupta & Sambrani (1978) while studying farming systems in north-eastern hills of India, concludes that, depending upon the agroclimatic conditions, a mix of activities, such as terracing, orchard plantations, forestry and forest-based industries, will have to be tolerated not merely because of emotional and socio-cultural factors but these activities provide the best economic opportunity in relatively remote and sparsely populated pockets. Further, an effective use of land will necessarily call for investment in related activities. Such investments should be considered as an integral part of the total programme and one must assess the projects on the basis of feasibility of the total scheme and not on the basis of single constituent unit alone. Moreover, the authors emphasized upon the creation of new department which within itself, comprises all these various competencies. A truly integrative effort needs to look at the proper match of resources and people.

Norman (1978) observed that farming system is the outcome of complex interactions amongst a number of inter-dependent sectors. So, coordinated and inter-disciplinary efforts are required for its improvement. Speeding (1979) reported that FSR is a group of interacting sectors, operating together for a common purpose. They are capable of reacting as a whole to external stimuli.

Patil (1978) observed after analysing the data collected from five villages of Veihe tehsil of Pune district of Maharashtra that crop production alone in itself is always risky and uncertain in the hill areas. To stabilise income in such areas inclusion of animal husbandry, dairy and forestry schemes and the infrastructural developments can substantially supplement the incomes of hill farmers. An economic analysis of the data collected revealed that there exists ample scope for exploiting dairy development activities because 71.27 percent of the total cropped area in the region is under natural grasses. He pointed out that the pattern of distribution of farm income from various sources is a good measure of ascertaining the type of a particular farming system. The study revealed supplementarity in the utilisation of main and by-products of one set of enterprise by another.

Bhati & Gopal Krishnan (1979) while identifying optimal farming systems for the western Himalayan region, revealed that the pace of population growth between 1951 and 1972 was more than the pace of growth in net sown area. They found a net increase of 609 animals per thousand hectares of uncultivated land. The authors suggested that exotic breeding should be intensified to improve the quality of livestock and the population of existing low yielding animals should be appropriately controlled. Further, the marginal and less productive lands devoted to cereals or pulses for meagre output is accentuating soil erosion. The author suggested that if marginal and sub-marginal land is utilized for raising orchards, forests or developed as pasture it can yield higher income. They further highlighted that in western Himalayan region, diversified farming systems which include crop husbandry, animal husbandry, orchards and forestry should be optimally developed in an appropriately integrated manner depending on the land capability. The hills and plains economies should be properly integrated and regional development plans based on the principles of comparative advantage, agro-climatic conditions and ecological considerations should be formulated and implemented.

Jodha (1979) studied dimensions of inter-cropping as practised in six semi-arid-tropical villages in three agro-climatic zones in Peninsular India and concludes that mixed inter-cropping is superior over sole cropping in terms of gross returns per hectare as well as per unit of peak period labour use. Its potential for greater employment is also revealed. The study further indicated that break- through in inter-cropping technology will help poor farmers more than the rich. Further the author stressed that higher resource allocation to inter-cropping will serve the equity goals better.

While discussing the farming systems in hilly areas, Shah (1979) noticed that cropping pattern in hills has not changed significantly with the passage of time. He observed that due to serious technical, economic, institutional and social constraints the potential has not been realised. The author suggested that farming system for the hills should include combination of different farming activities like cultivation of field and plantation of crops, animal husbandry, fisheries, sericulture, forestry, medicinal herbs, species of fodder grasses etc.

Singh, Patel and Ahlawat (1979) stressed that integrated crop and dairy plans have strong potential in increasing income and employment of small farms. Study of economic relationship between crop farming and dairying in developing areas showed optimal degree of complementarity between the two enterprises on different farm sizes. The results also revealed that the number of animals increased with farm size, but the number of animals per holding was low in cash crop areas then the traditional crop areas. The study further indicated that the combined profitability of the enterprises was found to be significantly higher on large farms with large size of herds.

Suryavanshi (1979) in his study of northern Maharashtra, observed that the rate of investment was higher on land followed by livestock and implements. The author noted negative relationship between farm size and proportion of investment on land. Food crops occupied more than 90 percent of cropped area. High yielding varieties were popular among almost every farm categories. In medium and large farms, the income from farming contributed 83 percent and 92 percent of the total income respectively.

Anandajaysekeram, Mtoi and Rugambisa (1981) and Reynolds and Lund (1981) emphasized that basic improvements are needed in the farming systems in order to improve the living standards of subsistence farmers. The farming community can only survive if new and appropriate farming systems are devised. These studies further stressed that a multi-disciplinary approach is relevant for solving the problems of food and energy.

Mansfield (1981) noticed that farmers behave rationally but the fast social and technical changes taking place in recent years have resulted in farmers adopting short term solutions which are not sustainable in the long term. The author emphasized the need for holding monitoring type surveys to update important farming system data, which are most susceptible to rapid change.

Desai (1982) pointed out that the main sectors of farming system are crops including horticulture, livestock, fisheries and forestry production. Smith (1986) stressed location specificity and farmer-oriented nature of farming system.

Singh & Singh (1982) conducted research in central Peninsular India and found that market differences in farming systems exist in the same locality. Adoption of improved cultivation practices in cropping systems, water conservation and water harvesting for supplementing irrigation, are some of the possibilities that need to be tested for technical and economic feasibility in the study area. Walker, Singh and Jodha (1983) suggested that crop diversification effectively imparts stability to net crop returns in the more rainfed farming villages. Authors suggested that for public investment in crop insurance, land consolidation and irrigation projects should be enhanced in order to stabilise farmers' income.

Bist (1984) studied the major constraints inhibiting the pace of agricultural development in the hilly areas of Uttar Pradesh [now Uttaranchal]. He noticed that the past experiences show complete lack of proper integration between related departments and programmes. Agricultural development efforts did not make a breakthrough due to segregative attitude of development agencies. Looking into achievements of these agencies the author proposed that the principle of agricultural development must incorporate dominant constraints such as topography, climate, land-man ratio, net sown area together with ecological planning principles. He further stressed the need of integration between soil conservation, livestock, horticultural, forestry programmes along with crop husbandry programmes for the overall development of the hill farms. He suggested that the action plan demands a drastic change to bring agriculture, horticulture, animal husbandry, and forest, under a single line of administration. The new economic activities like wasteland development programme, community land development programmes, horticultural activities, off-season vegetable etc. can make a dent in the regional development. These activities offer some of the viable lines of economic activities which strengthen the fragile economic base of the hill areas by providing employment potential to rural masses, alongwith sufficient income provisions.

Norman & Collinson (1985) stressed over the heterogeneity that exists at the farm level and which also appears to be growing with the passage of time. Bearing in mind the farm level differences in technical and human elements, heterogeneity must be explicitly recognised and strategies be designed to help the farmers. The authors strongly suggest the adoption of farming system research approach to future research. The F.S.R. can be made effective if there are strong two-way linkages between farmers, agricultural scientists, technical and social and policy making/implementing agencies.

Reddy & Korwar (1985) studied comparative economic evaluation of agro-forestry, silvo-agriculture and silvo-pastoral systems and found that annuity value worked out to Rs. 799 in case of agro-forestry, Rs. 598 in case of silvo-agriculture and Rs. 1142 in silvo-pastoral system. The corresponding cost-benefit ratios were 1.45, 1.29 and 2.25. Agro-forestry and silvo-agriculture provided foodgrains apart from firewood and fodder respectively. On the other hand, silvo-pastoral system helped in producing nutritious commodities such as milk/mutton. The study suggested that these systems are to be practised according to the availability of land resources and farmers' needs. Besides this, these systems would yield intangible benefits like control of soil erosion, stabilisation of prices of foodgrains/fuel and improvement in the environment.

Tewari (1985) analysed the economics of various agro-forestry systems after introduction of subabul plants in the existing cropping system of lower hills of Himachal Pradesh. The study revealed that per hectare net returns have increased nearly three times from Rs. 598.62 in the existing cropping system to Rs. 1350.69 with the adoption of agri-silvi-pastoral system. The integration of trees with crops/pastures was found to be more economical than cultivation of pure agricultural and horticultural crops.

Jodha (1986a) observed that the stagnation in production and uncertainty in dryland farming systems was due to neglect in the development of new technology. He stressed the need for systematic wholesome farm research based on the type and climate rather than individual crops. Sampath, *et al.* (1987) found that sequential and inter-cropping system should be used to increase the efficiency of renewable and non-renewable resources.

Mahat (1986) analysed the need for integrated management of resources in mountain watersheds by studying the inter-relationship of forestry with agriculture and livestock husbandry. He identified decreasing land man ratio and decreasing farm productivity as the main constraints for under-development of hill economy of the Hindu-Kush Himalaya. Farm productivity in the hills is declining due to the high level of its dependence on forest, low forest to agricultural land ratio and declining forest productivity. For increasing the farm incomes, hill farmers try to make up production deficit by augmenting their cash income through extension of agricultural land to marginal areas, intensification of agriculture, adoption of advance technology and resorting to off-farm employment. The adoption of new technology is constrained by farmers' economic inability to buy critical inputs. The characteristic features related to employment and income structure are high dependence proportion of population on agriculture, scarcity of skills, low level of industrial development and limited employment opportunities. The author suggested that the promotion of farm forestry through appropriate activities and investments can generate income and off-farm employment opportunities for the rural people in the hills.

Oberoi & Moorti (1986) worked out the economies of sheep farming in Bharmaur tehsil of Himachal Pradesh. The study indicates that sheep enterprise is a continuous source of income, by providing three clips in a year. The share contributed by that wool in the total income is 9, 19 and 59 percent on small, medium and large farms respectively.

Farrington & Martin (1987) stressed the role of economies in the farming system research and stated that it should be an applied problem solving dynamic and interactive approach, based on the scope and potential impact of changing technology within holistic farming system framework by identification of relatively homogenous group of farmers within a zone as clients of research.

Krishnamurthy, *et al.* (1987) revealed that the livestock population is India is increasing by 2.5 percent per annum and the fodder requirement would be around 145 metric tonne against the present production of 100 metric tonne. In order to increase the total biomass production from the unproductive lands, they stressed the alternate crop-based systems. Pathak and Roy (1987) pointed out that wasteland under the forest area combined with silvi-pastures programmes have a potential of meeting half of the national forage needs.

Parthasarthy & Satyanarayan (1987) found that the integrated approach for horticultural development in Andhra Pradesh has resulted in increase in horticulture crops. The future thrust on research warranted dryland horticulture development, crop improvement by hybridisation, eco-physiology of tropical fruit crops, crop nutrition, cropping systems, water management practices, postharvest technology, forestry and aromatic and medicinal crops. The integrated approach involves supply of pedigree plant material, training persons to equip them with technical and managerial capabilities, establishment of horticultural estates to achieve integration of production, processing, marketing and identification of locations to set up industries based on horticultural resources.

Paris & Duff (1988) discussed use of gender analysis based on women relation information in the context of whole farming systems in the two research sites in Philippines. The study concluded that by incorporating women's concern into the farming system research process could lead to a deeper understanding of the farming systems.

Tothill (1988) observed the presence of strong linkages between livestock and crop production on small farms in highland of Ethiopia. He stated that linkages were high in mixed farming system because of consumption of leguminous forage crops by the animals on the one hand and prevent decline in soil fertility. Therefore, he suggested the introduction of leguminous crops in each crop rotation. This will be helpful to stabilise the system in many ways, such as animal traction and nutrient recycling.

Salam, *et al.* (1990) examined the economies of homestead agricultural farm in Kerala state. The farm was categorised by the nature of cropping, livestock and irrigation system. The study concludes that net income of Rs. 1460 was generated after meeting all annual family expenditures. It is further suggested that this farming system is a sustainable and profitable for small farmers and therefore, should be encouraged at large scale.

Joshi & Tyagi (1991) conducted their study on existing farming systems in Punjab & Haryana during pre- and post-green revolution period and observed that the rate of production and yield of major crops had declined in the states. They also concluded that the existing farming systems are unsustainable in the long run.

Sharma, Bhati and Singh (1991) in their study, conducted in four different zones of Himachal Pradesh found that in low-hill zone remittance-supported crop farming system is existing. In mid-hill zone two distinct farming systems have emerged: (a) livestock-based farming system, and (b) vegetable-based farming system. In the high-hills both in dry and wet-zone fruit-based farming system is emerging. Authors found that all the three systems are depicting the system of long run unsustainability. The study also pointed that there is a strong inter-dependence among different sub-systems of farming in each zone of Himachal.

Bhati, Rathore and Sharma (1992) found that there is a great diversity in mountain farming system in different zones of Himachal Pradesh which have emerged in response to and adaptation to peculiar agro-climatic environment. According to the study, there is great diversity in level and composition of rural household income in different zones of Himachal Pradesh. Although family size and farm size have not differed much in different agro-climatic zones but there was a wide variation in farm household income due to qualitative differences in farming systems.

Chatterjee (1992) noted that complete farming system approach is a synergistic and catalytic in nature. In this system each component extract strength from the other to create self-perpetuating multiplier effect on the economy. For the success of the system, he stressed optimum use of scarce resources for the betterment of poor and small farmers.

Bhati & Sharma (1996) concluded that there are wide variations and a high degree of heterogeneity within the hilly regions regarding physical conditions, soil characteristics and micro-climates which result in evolution of different farming systems in different sub-regions of H.P. Therefore, in formulating a micro-level economic development strategy, farming system approach should be followed for integrating food sub-component, fodder sub-component, fuel sub-component and off-farm income generation components of the farming system into one integrated plan.

Lipton (1997) mentioned that in terms of employment the number of people employed, agriculture is the most important single industry (2.5 billion) in the world. So, it is more important to grow agricultural productivity for the general welfare of the populace. This is particularly true for Africa where food problem per person actually fell over the 20 years, but argument may be extended to other parts of "tropical world". The author also suggests reduction of imports and great self-sufficiency in food should yield benefits through import substitution and saving of scarce foreign exchange. Author considers agricultural development as a necessary condition for the growth of industry. So, proper use of labour, natural resources and capital services are required to solve the problem of food security in the developing world.

Contribution of Common Property Resources

Common property resources have received attention from research scholars, environmental economists and natural resource specialists in recent times because of crucial role played by these resources in farm and household activities. These resources are treated as free goods by society, planners and policy-makers and have remained outside the domain of the market. Consequent productivity of these resources has declined due to over-exploitation, encroachment and poor management. Important studies conducted in the field of CRPs in India and abroad are reviewed hereunder:

Fischer & Krutilla (1974) defined CPRs as resources not necessarily owned in common by all members of the community where neither exclusion nor discrimination is permitted in respect of their access. Bromley and Cernea (1989) defined common property resources as those resources in which all the people have equal use rights. According to them, the membership in a group of co-owners is conferred by membership in some other group, generally a group whose main purpose is not the use or administration of the CPRs, e.g. a village or a tribe. The similar view regarding common property resources was given by Jodha (1986b). According to him common property resources are those resources which are accessible to the whole community of a village and no individual has exclusive property rights.

Maheshwari, *et al.* (1981) studied the consumption pattern of fuels in a village of Bhopal district, and concluded that per capita consumption of fuelwood was of 3.1 quintal, 3.2 quintal of dung cake, 5.1 litre of kerosene and 21 kilowatt per hour of electricity by the rural houscholds of the study area. Among the energy resources diesel and electricity were mostly used in agriculture and it was found that domestic sector was the major consumer of energy (84%).

Hartwick & Olewiler (1986) have pointed out that the land under common property ownership is over-exploited as compared to private lands which has been demonstrated through a case study of wheat production by varying the level of labour by using quadratic production function.

Levine, *et al.* (1986) has observed that CPRs play a significant role in the life of poor people in rural India. They also mentioned that the development projects involving common lands and biological resources encounter a number of problems relating to the equitable sharing and access, maintenance and technology which are significantly different from private lands.

Damodaran (1988) studied the emergence of grazing crisis faced by sedentary village communities of India which own the animals. He found that common pastures and other grasslands were deteriorating and severe undernurishment of livestock was identified as major facet of the crisis.

Sexena (1988) analysed the role played by the grazing lands in the well-being of the desert people and observed that degradation of grazing land was mainly due to high pressure of cattle grazing. Measures suggested for improving these marginally deteriorated lands by the author are protection from biota, soil conservation, re-generating grass species. According to Gibbs, *et al.* (1989) common property rights are special types of rights which assure individuals to access a resource over which they have a collective claims. Common property regimes are form of management grounded in a set of accepted social norms and rules for the sustainable and inter-dependent use of collective goods such as forest, grazing grounds, fisheries and water resources. They found that common property regimes can be efficient and equitable mechanism for sustainable resource management and may provide an appropriate foundation for their adaptation of institutional arrangements in the future.

Wade (1988) revealed that some villages also develop common property resources of grazing and irrigation, but others do not. It was observed that the villages located towards the tail end of the irrigation system have large corporate organisation than elsewhere. He has found that collective management of CPRs can be an alternative to the privatisation and state control to arrest the degradation.

Pickup (1989) noticed in his study that Australia's arid grazing lands have eroded and degraded by trampling during the domestic stock grazing resulting in changed botanical composition of pastures and the changed composition interact in space and overtime and reducing productivity of the system. The author suggested/felt the need to use new survey methods (remote sensing), which can separate grazing induced degradation from that of natural variability.

Chopra, *et al.* (1990) emphasized that people's participation in the management of common property resources will increase the productivity of the privately-owned assets in lower Shiwalik ranges of the Himalayas.

Moorti, *et al.* (1990) found that about 60 percent of farm income has been contributed by livestock enterprise. They have observed that the fodder scarcity, land closures by forest department, non-availability of winter grazing lands and lack of awareness about afforestation is causing hardship to the shepherd community. The study about costs, availability of water, resource use efficiency and water distribution under different sources of irrigations by Sharma (1990) in Kangra district of H.P. has found that kuhls are the cheapest source of irrigation.

Gupta, *et al.* (1992) studied the status of fodder and fuelwood in the states of Himachal Pradesh and Jammu & Kashmir. The study found that the share of grasses to total available fodder was maximum followed by straw in both the states. Firewood is the main source of domestic energy. Sixty to seventy percent requirement is fulfilled from the farm resources of the farmers and remaining is supplied by the public lands. The dependency on fuelwood was relatively low as compared to the fodder.

Garen (1993) found that most of the CPRs in India have been degraded due to their conversion into open access, increasing pressure due to growing number of extractive industries, commercialisation of rural sector has led to the breakdown of conventional management practices. Further, it was noted that privatisation, social forestry programmes, tree growers, cooperatives and non-governmental organisations are the success stories in some cases.

Chand (1994) stated that the kuhls are the mainstay of irrigated agriculture in Himachal Pradesh. The farmers had water rights from erstwhile rulers and later on were confirmed by the state at the time of land settlement. This has encouraged the water users to form organisations generally called as Kohli/Kuhl Committee. Government's support in the form of financial aid for repair, maintenance, erection of kuhls etc. has made the farmers dependent on government for the kuhl management as well.

Baker (1995) has conducted a study to examine the kuhl management in Kangra district of H.P. and found that radical changes have occurred in these regimes in the form of water scarcity and poor cooperation among the members. This was due to the lack of interest of the kuhl members, better employment opportunities, changes in socio-economic factors and farmers are shifting towards low water intensive crops.

Singh & Bhati (1995) revealed that in all categories of rural households of Himachal Pradesh, firewood is the main source of domestic energy available from public and common forests. The dependence on common pool resources (CPRs) was found to be the highest among marginal farmers.

Singh & Dhillon (1995) conducted their study in Kandi area of Punjab to study the impact of developmental works on income and employment. They further found that the proportion of households grazing their livestock on common lands has decreased from 60 to 58 percent, while the share of green fodder available on the farms has increased from 58.4 to 60.7 percent.

Management Strategies for Common Property Resources

Gupta (1986) forwarded the view that technological solutions to low productivity of grazing lands of arid and semi-arid regions are limited. In various management options like privatisation and closure of common grazing lands, all users must be treated equally. This affects the landless livestock farmers more adversely.

Frequent droughts in these regions have impaired the ability of small farmers and agricultural labourers to adjust to the lean periods through livestock management. Author has recommended socio-ecological approach to the problem of grazing land management in these regions to solve large scale tensions.

Agarwal, (1988) mentioned that north -western states of India have large proportion of tribal population living in and around the forest on which they depend almost entirely for their subsistence. They practice shifting cultivation and collect variety of products like fuelwood, fruits, fodder, medicinal plants etc. Forestry operations also provide employment for the people. The study suggested that suitable village organisation can promote the afforestation of village common lands and on private lands on mutually agreed terms with proper regard for rights and privileges of both the inhabitants and administration.

Singh & Ballabh (1989) overviewed the afforestation of village common lands and found that village wood lot was well planned and administered by forest department. The village panchayat received an adequate annual return from this wood lot and the villagers were supplied free fuelwood and fodder. It was suggested that the government intervention in the management of common lands can avert the tragedy of commons. The villagers' participation in such schemes can augment the success of these programmes.

Singh & Singh (1989) analysed the hill farming system of Dwarahat block of Almora district. Authors found that the productivity in terms of biomass (cattle fodder, fuelwood, timber, etc.) has declined and the system is no longer able to support the human and livestock population. They have suggested the rehabilitation of village common lands through afforestation, use of more efficient stoves, improved composting techniques and other measures to increase the crop productivity.

Chopra & Kadekodi (1991) analysed the concept and compared the working of participatory institutions in rural India. In a two part study, the first part analysed the management of community lands, water, forest and other common property resources entrusted to the village community. The second part analysed the management of CPRs created by pooling of private property lands with marginal productivity. The study concluded that under certain set of conditions, both processes lead to development that conserves environmental resources, such as soil, water and biomass.

Frederick & Sedjo (1991) made an historical examination of the use and management of the USA's water, forest, range land, soil and cropland and wild life resources in the context of the growing role of outdoor recreation. He argued that sustainability is not possible without management and management is not possible without establishing institutions which provide economic incentives for producing or conserving resources. The fact of the matter is that many of these resources which heighten the importance of establishing institutions will lead to a socially optimal resource development.

Gilmour & Nurse (1991) compared the tree cover in Jhiku Khola catchment area near Kathmandu through photographs taken during 1972 and 1989. And it showed that the percentage of major agricultural land use category and common lands have increased significantly from 1972 to 1989. Indications show that overall landscape is maintained well by the tree-based farming systems in the mid hills of Nepal.

Thakur, *et al.* (1995) found that sheep and goats are an integral part of farm household system in the fragile and topographically different areas of Himachal Pradesh. Sheep and goat-rearing was found to be a sustainable and highly profitable enterprise. They have suggested that in order to maintain a balance between ecology, environment and sheep and goat-rearing there is a need to expedite and monitor some of the measures like improvement in pasture management, and plantation of fodder trees in order to maintain balance between ecology, environment and sheep and goat-rearing.

CHAPTER THREE

The Study Area

The present chapter is devoted to discuss the procedure followed for sampling design to reach the final respondent, who form the basis of the study and to highlight the tools adopted etc. In the present investigation, emphasis is laid on farming system analysis and common property resources (CPRs).

Selection of the Study Area

The study area broadly comprised of state of Himachal Pradesh which is situated in the lap of Himalayan ranges in the extreme north-west of India. The state of Himachal Pradesh is prominent in the western Himalayan region for reasons of its being ahead of other hill areas, specially in the field of agricultural development. The state has a deeply fissured topography, with mountain ranges and valleys of varying slopes and sizes and low population density. Physical characteristics of the state highly restrict the amount of cultivable area. Forest and water resources are two prominent natural resources. The land resource is very limited, the net sown area being only one tenth of the geographical area. Average size of holding in the state is only 1.21 hectares (1997-98) which by any standards can not be considered economical. This is particularly so when holdings are fragmented, terraced, soil is stony and lacks desired quality, and irrigation is inadequate. The state being hilly, the altitude varies from 350 ft. to more than 7000 ft. above MSL.

The present study is aimed at studying the different farming systems present in the state and their detailed aspects. Thus, the study had to be conducted in such an area where largest number of farming systems were present. Moreover, this area had to be contiguous and secondary data should have been available for it. Thus, the choice was to be based on administrative units and hence one district was to be selected for the purpose. The search for such a district stopped at district Kinnaur because of the availability of different farming systems, e.g. agricultural, horticultural, livestock-based etc. All these systems are interlinked and CPRs has main role in this zone. Thus, district Kinnaur was purposely selected for the study.

District Kinnaur comprises of three development blocks viz. Pooh, Kalpa and Nichar. These account for 29.84, 36.67 and 33.47 percent respectively of the population of the district. After selecting the district, the next stage of sampling procedure was the selection of blocks. Due to variation in vegetation types and geographic and climatic features, no single block can reflect the

Table 3.1. Nucleus villages and the study villages

Zone	District	Nucleus village	Other villages	Total No. of villages
Dry temperate zone	Kinnaur	Pooh	Nako, Leo, Maling, Doubling, Doobling, Ropa, Gybang, Sunam, Kanam, Spillow, Moorang & Thangi.	12
		Kalpa	Chini, Kothi, Powari, Purwani, Sangla, Sapni, Kilba, Chhitkul, Raksham & Ribba.	10
		Nichar	Sungra, Tranda, Beri, Kafnoo, Yungpa-I, Katgaon, Chigoan, Tapri, Urni, Rupi & Chhota Khamba.	11
			Total	33

clear picture of district Kinnaur. In order to give wide geographical coverage in the study all the three blocks were selected for the study.

In the next stage, a complete list of revenue villages in each of the blocks was drawn. Thus, the lists of 80 villages in Pooh, 63 villages in Kalpa and 85 villages in Nichar were drawn. One village from each of the block was purposely selected as a nucleus village, taking into an account the characteristics of the zone. Around each nucleus village 12 villages in Pooh, 10 villages in Kalpa and 11 villages in Nichar were selected to form the clusters of sampled villages. Table 3.1 presents the information regarding the three nucleus villages for the selected blocks alongwith the other study villages.

Selection of Households

A preliminary field visit was made to get an idea or overview of the farming systems, customs, traditions, historical background and institutional arrangement etc. A schedule was developed on the bases of this visit.

The patwaris of sampled villages were contacted and lists of households with the size of holdings were drawn. From each cluster, a complete list of households was prepared. The cultivators were selected randomly based on probability proportion to their numbers in each category.

Data were collected from these finally selected households on a pre-designed and pre-tested schedules. However, during actual data collection many farmers had to be dropped because of their non-cooperation and also during scrutiny of schedules a few more schedules had to be discarded due to inconsistency of data. Thus, the final sample consists of 225 households instead of originally planned 255 households. The distribution of the final sample is presented in Table 3.2.

The above classification of farmers into marginal, small and medium category is based on standard farm size classification.

The farmers who had holdings upto 1 hactare were termed as marginal, 1-2 hactare were classified as small and 2-4 hactare land holding farmers were termed as medium farmers. The number of large farmers in the sampled villages was found negligible, hence this category was not taken into consideration.

Table 3.2. Block and distribution of sample households (category-wise)

Block	Category			Total
	Marginal	Small	Medium	
Pooh	27	35	6	68 (29.84)
Kalpa	31	37	15	83 (36.68)
Nichar	23	40	11	74 (33.48)
Total	81 (36)	112 (50)	32 (14)	225 (100.00)

Note: Figures in parentheses are the percentages.

Data Collection

Both primary as well as secondary data were collected to fulfil the requirements of various objectives of the study.

Primary Data

Survey method was used for collection of data on specifically prepared and pre-tested schedule by contacting the farmers according to their availability in fields or in their houses. Primary data collected include information of family size, occupation, education level, tools & equipments, cropping pattern, livestock activities, forest inventories, off-farm activities, farm inputs and outputs with their respective prices, problems faced by farmers, credit structure, fuelwood and timber requirement, consumption pattern, farm forestry, irrigation, etc.

The information regarding the fuelwood and fodder collection, animal litter, grazing etc. in forests and village common lands were obtained from the sampled households. The detailed information concerning the kuhl irrigation system, organisational setup, water distribution system, kuhl management, repairs and maintenance were collected. The problems and suggestions of the respondents in this regard were also recorded for the better management and sustainable use of these resources.

Secondary Data

Secondary data were collected from various sources and state agencies like Department of Agriculture, Department of Horticulture, Directorate of Land Records, Directorate of Economics and Statistics, Forest Department and various other government offices. Various published and unpublished documents were also consulted and required data were sifted.

Analytical Methods

There are various tools and techniques available to researchers for analysis and arriving at conclusions. However, in the present study the method of simple tabular analysis has been adopted because of its simplicity and also because the results are straight forward. Tabular method was used to study the socio-economic features, identification of farming systems and analysis of the responses of the farmers regarding the problems in resource use and their suggestions. The conclusions have been drawn by working out and analysing the averages, percentages and indices etc. as under.

$$\text{Sex ratio} = \frac{\text{Total number of females}}{\text{Total number of males}} \times 1000$$

$$\text{Literacy rate} = \frac{\text{Total number of literate persons}}{\text{Total population}} \times 100$$

$$\text{Literacy index} = \frac{\sum W_i X_i}{\sum X_i}$$

W_i = Weights 0, 1, 2, 3, 4 and 5 assigned to illiterate, primary, middle, matric, graduate and above.

X_i = Number of persons falling in the ith educational category.

$\sum W_i X_i$ = weighted total of literate population

$\sum X_i$ = Total population

$$\text{Cropping intensity} = \frac{\text{Cropped area}}{\text{Net area sown}} \times 100$$

Input-Output Model

Static input-output model has been used to study the inter-linkages between the different farm components. The inter-linkages have been worked out for the seven sectors, using the micro level data from different sectors. The sectors considered for the model were agriculture, livestock, horticulture, farm forestry, common property resources, labour and the market. In this model, household sector is the focal point which controls all the contributing sectors for the inputs required and outputs generated by various farm components under different farming systems. The basic form of the input-output model is given below:

$$x_i = x_{ij} + F_i \quad \text{.....................................} \quad \text{I}$$

X_i is the value of output from i^{th} sector and xij and F_i are the amounts of X_i used in household (F_i) and other sectors of the system (j, s), j = i, 2,...n).

$$X_i = \sum X_{ij} = F_i \quad \text{.................................} \quad \text{II}$$

The equation I can be expressed as transaction matrix. The transaction matrix shows the value of inputs that flows from one sector to other. In this way the contribution of all the sectors can be expressed as the total cost of production for the sector and associated net income from that particular sector. The coefficients, thus obtained can be expressed as input-output coefficients and may be denoted as

$$a_{ij} = \frac{X_{ij}}{X_i} \qquad \text{.......................................} \quad \text{III}$$

$$X_{ij} = a_{ij} X_j \qquad \text{.......................................} \quad \text{IV}$$

In the above formulation every a_{ij} gives the rupee's worth of the produce of the ith sector required by j^{th} for per unit value of output of jth sector.

$$X_i = åa_{ij} X_i = F_{ij} \qquad \text{..........................} \quad \text{V}$$

This represents the relationship existing between the household sector and output of X_i and the relationship between intermediate sectors a_{ij} in the farm.

Valuation of Inputs and Outputs

The inputs like fertilizers, seeds, insecticides, pesticides, etc. have been valued at actual expenditure incurred by taking the market price into consideration. For the valuation of family labour, FYM, own seed, services of bullock labour (owned and hired), rates prevailing in the locality were used. Similarly for valuation of the farm produce kept for home consumption, local prices were taken into consideration. In case of marketed produce sale prices were taken into account. While aggregating the income of households from different sources care has been taken to avoid double counting. Gross income from crop husbandry have been calculated by multiplying the physical quantities of different crops by post-harvest prices of both main and by-products. Gross farm income from livestock comprises value of milk, receipts from the sale of animals, value of farm yard manure, value of animal power used/sold etc.

The imputed income from forest produce like timber, fuelwood, fencing materials, leaf manure, fodder, grass and free grazing, litter etc. has been assessed separately for farm land forestry and common property resources. The imputed income from common property resources was calculated. Practical difficulties were encountered while estimating the value of fodder consumption by livestock during grazing. However, keeping in view the average number of animal per household the total monetary worth grazed fodder was estimated. The imputed valuation of forest produce extracted by households from common property resources is used to estimate total income.

Contribution of CPRs

The contribution of forest and common grazing lands to the households has been worked out by estimating the quantity of fodder saved. It was estimated by taking the differences in grass fed to different farm animals under stall-fed and grazed conditions. The estimated quantity was valued at the prices prevailing in the locality for the grasses. Similarly, other contributions of the forests viz., fuelwood, fodder, timber etc. were estimated and valued at the prices prevailing in the locality for the similar farm produced outputs.

The additional income generated by kuhl irrigation was estimated because irrigation is the part and parcel of Kinnaur district. Various problems related to kuhl irrigation were also observed.

Limitations of the Study

Despite best intentions the study suffered from the following limitations:

1. Some of the respondents due to ignorance were apprehensive of the intentions of researcher and despite best efforts did not cooperate in imparting data.
2. The accuracy of data was also impaired to certain extent by memory bias of the respondents.
3. Clear cut breakup of the area under various agricultural and horticultural crops was not available as farmers tend to raise these crops as joint crops/not on a distinct plot.

CHAPTER FOUR

Socio-Economic Features of Sample Households

Before presenting results of the analysis of the problem under study, it will be useful to discuss the socio-economic background of the sampled respondents. The present chapter is a step in this direction. It is sometimes assumed that the socio-economic profile is in no way directly related to the actual problem under study and hence, is redundant. However, it may be mentioned here that the socio-economic profile presents the conditions under which the sample is operating and the results of the study can be viewed under this light. Moreover, while comparing the result of one study with another carried out in some other place at some different point of time, the socio-economic profile provides the basis for comparison. As such, it is considered important to touch this aspect of the sampled households and has been presented as follows:

Family Size

The social characteristics of the sampled households have been presented in Table 4.1.* The table reveals that the average family size is 6.67 persons. Direct relationship of family size with holding size is also observed. It increases from 5.24 in case of marginal farmers to 10.25 in case of medium farmers. The same pattern is evident in all the blocks under consideration. The block-wise scrutiny further reveals that highest family size is in Nichar block

*See Tables 4.1 to 4.29 at the end of this chapter.

(7.07) followed by Kalpa block (6.73) and Pooh block (6.15). The family size of Pooh and Kalpa blocks though almost identical differ substantially among individual size classes.

Religion

Hinduism and Buddhism are two prevalent religions of the area under study. The Hindu population accounts for 80.44 percent of the total population in the area and the share of Buddhist population is 19.56 percent. The Hindu population is dominant in all the three blocks. In Pooh block 61.77 percent are Hindus and 38.23 percent Buddhists whereas in Kalpa the percentage of Hindus increases to 87.95 percent and there are only 12.05 percent Buddhists in this block. The percentage of Hindus further increases to 89.19 percent in Nichar and only 10.81 percent Buddhists are observed there. The Pooh block has boundary with Tibet, therefore, impact of Buddhism is more as compared to the other two blocks.

CASTE

The caste-wise breakup of the population shows that Scheduled Castes constitute 32.0 percent and 68 percent belong to general castes in the study area. In Pooh block about 27.93 percent persons belong to Scheduled Castes whereas these percentages are 38.54 percent in Kalpa and 28.38 percent in Nichar block.

Family Type

The details of family structure reveals that 74 percent of the total families are of nuclear (Table 4.1) and 26 percent are joint families. The figures show that percentage of nucleus families are high in Pooh block (79.41%) followed by Kalpa block (74.27%) and Nichar block (68.92%). This indicates that trend is towards nucleus type of families in the selected blocks.

MARITAL STATUS

The marital status of the persons in the sampled households has also been included in Table 4.1 indicating predominantly large percentage of married persons. At overall level about 96 percent persons are married and the higher percentage of married persons is found in Nichar (97.30%) followed by Kalpa and Pooh having 96.39 and 92.64 percent respectively. The cent percent sample of medium farmers of Pooh and Kalpa and marginal and medium farmers of Nichar are married.

Educational Level

Educational level of the population is an important indicator of development. The higher levels of education are always desirable for the society and considering its importance the educational levels of males and females are analysed and presented separately in details.

Male Education

The educational levels achieved by male members of sample households have been presented in Table 4.2. The table reveals the very high level of percentage of literate among the male members and at overall level about 90.47 percent males are literate whereas there are only 9.53 percent who are unable to read and write even a simple letter. The higher level of literates is found in Nichar block where about 94 percent males are literate and this is followed by Pooh having 90.49 percent and Kalpa having 87.70 percent male members literate. The observation reveals that highest percentage of literate is among small category of farmers at overall level. Almost same trend is present in individual blocks. In addition to studying just literates or illiterates, the analysis is extended further to include the formal educational levels as well. There are 22.63 percent males who have passed primary school, 30.69 percent have passed matriculation examination and 18.26 percent graduates and 3.45 percent postgraduates are also included in the sample.

Female Education

The female educational levels achieved have been presented in

Table 4.3. It may be seen from the table that in district Kinnaur about 83.46 percent females are literate, which is a commendable achievement. However, this is about 7 percent less than the level achieved by males. The analysis further reveals that the female literacy increases with holding size in the study area. In case of marginal farmers this percentage is 76.02 which increased to 85.94 and 93.66 percent among small and medium farmers. At block level the lowest female literacy is in Nichar block at 82.78 percent which increases to 84.09 percent in Kalpa and 83.45 percent in Pooh block. At overall level 20.57 percent females are only able to read and write a simple letter, about 25.19 percent females have passed the primary classes and 27.28 percent have passed matriculation examination. The sample also includes 8.41 percent female graduates and 2.02 percent female postgraduates. The comparative analysis reveals that female levels of educational achievement at matriculation and above are lesser than their male counterparts.

When analysis is joined together in this respect the picture emerging has been presented in Table 4.4 and it is clear from the table that 86.99 percent persons are literate in district Kinnaur whereas in entire sample a percentage of only 13.01 being illiterate. The figures clearly indicate that the vast majority of population being literate, must be open to known ideas of social and economic development. As such it is very easy to motivate the farmers to adopt new production techniques on more efficient cropping pattern, land use pattern etc. The block wise analysis of total literacy has also been presented in this table. In Pooh block the percentage of literates is 87.06 percent whereas this is 85.85 percent and 88.19 percent in Kalpa and Nichar blocks respectively. The analysis also indicates that as the holding size increases the percentage of literacy also increases. May be the higher income from large holding and higher awareness among large land holders, were the motivating factors behind this scenario.

Occupational Structure

It is generally observed that majority of rural people have agriculture or horticulture as their occupation. But due to seasonal nature of these activities and the fact that small holdings don't consume all the time available with farmers, many persons also engage themselves in some other economic activities. These activities consume smaller proportion of their time and share of these in total income is also less. These occupations are termed as secondary occupation whereas the former is called the primary occupation. In the present work both of these have been analysed and the results are presented in Table 4.5. The results are interpreted separately for main and secondary occupations.

Primary Occupation

The main production activity of farmers in district Kinnaur is horticulture, due to conducive agro-climatic conditions. This is also reflected in the analysis that in Kinnaur 51.56 percent persons have horticulture as their primary occupation followed by agriculture which is primary occupation of 40 percent persons. Next in importance is the services, occupying 5.34 percent persons followed by livestock rearing (1.78%), trade (0.89%) and non-farm labour (0.44%). The trend among other occupations is almost similar except that no person has trade as his main occupation. Even higher percentage of persons in Pooh has horticulture as main occupation (55.88%), 35.29 percent have adopted agriculture as their main source of livelihood and 5.88 percent are engaged in government services. But trade is not observed as an occupation in this block. The number of persons adopting horticulture as primary occupation dramatically fell to 39.14 percent in Nichar block where 50 percent persons have agriculture as primary occupation. No non-farm labour and person engaged in trade is found in this block (Table 4.5).

Secondary Occupation

The pattern of secondary occupation is found to be different than the primary occupation. Agriculture is the most important secondary occupation; 44 percent persons are engaged in it and this is followed by horticulture, the secondary occupation of 22.14 percent persons. Large number of persons have adopted livestock-rearing as secondary occupation and 17.79 percent persons being engaged in it. Services occupy 11.61 percent and trade 4.48 percent persons. No person is found to be working as non-farm labour as secondary occupation. The picture in Pooh and Kalpa blocks is almost similar where about 50 percent persons have agriculture as their secondary occupation and about 20 percent are engaged in horticulture. This distribution is more equal in Nichar block where 29.72, 25.44 and 20.27 percent persons are engaged in agriculture, horticulture and livestock-rearing respectively (Table 4.6).

Land Utilisation Pattern

Land utilisation pattern of a village, block or district or for that matter of an individual indicate the efficiency by which this scarce service is being utilised. There is a possibility of increasing the farmers' income by adjusting the land use pattern towards more efficient one. The existing land used pattern of the sample has been presented in Table 4.7 separately for each block in district Kinnaur.

The table reveals that in Kinnaur each farmer has 20.48 bighas of land; out of this about 60.04 percent is irrigated and the rest about 39.96 percent can be categorised as unirrigated . The highest percentage of irrigated land is with medium farmers whose about 65.81 percent of holding is irrigated. This is followed by marginal farmers who own about 61.98 percent of land which is irrigated. The last in this respect are small farmers whose only 59 percent of land holding is irrigated. The block-wise analysis reveals that in Pooh and Kalpa blocks about 84 percent of land is irrigated. But in Nichar only 11.44 percent of land is irrigated and large portion of 88.56 percent is unirrigated. The marginal farmers of Pooh, medium farmers of Kalpa and small farmers in Nichar owned largest proportion of irrigated land in their respective blocks.

Further analysis reveals that the maximum proportion of land, at district level, is under the cultivation. The orchards are in popular use accounting for 37 percent of total cropped area. Some part of land is also under forests, pastures and ghasnies. More or less in Pooh block there is no land under forests.

Cropping Pattern

Cropping pattern is the study of types of crops grown by the farmers and the area utilised for this purpose. The cropping pattern at the block level and for the district as a whole has been presented in Tables 4.8 to 4.10 with the area under different crops and the resultant production. The details of each of the crop has been presented in the following text.

Maize

Maize is grown by all categories of farmers in all the blocks. In Pooh and Kalpa blocks this crop is grown completely on irrigated land, whereas in Nichar only about 19.37 percent of area under maize is irrigated. At overall level, each farmer, on an average, devoted 8 percent of cropped area for cultivation of maize of which 73.48 percent is irrigated resulting in 1.19 qtls. of production. In Pooh, on an average 8 percent cropped area produces only 0.77 qtls. of maize. But in Kalpa farmers are able to produce 1.10 qtls. of maize from only 7 percent of land. These figures for Nichar are again 8 percent and 1.66 qtls. respectively. It is also seen that irrigation facilities do not influence the productivity of this crop. The production of maize is highest in Nichar block (0.88 qtls./bigha) followed by Kalpa (0.57 qtls./bigha) and Pooh (0.42 qtls./bigha) respectively. But in district Kinnaur as a whole, productivity is 0.63 qtls./bigha.

Miscellaneous Crops

Miscellaneous crops include ogla, phapra, koda, chollai, bathu, mash, kulth etc. These crops are considered to be inferior crops but are popular crops of Kinnaur. Each farmer, on an average, is devoting only 5 percent of his land for this crop and produces 0.85 qtls of output. About 75.02 percent of land under this crop is irrigated. In Pooh and Kalpa blocks all the land under this crop is irrigated whereas in Nichar the percentage of irrigated land under this crop is only 24.03. In Pooh nearly 4 percent of land is under miscellaneous crops with 1.02 qtls. of output. The comparative figures for Kalpa and Nichar blocks are 5 percent and 0.77 qtls. and 6 percent and 0.87 qtls. respectively. In aggregate the productivity of these crops is highest in Pooh block (1.00 qtls./bigha) followed by Nichar block (0.66 qtls./bigha) and Kalpa block (0.58 qtls./bigha) respectively.

Rajmash

In Kinnaur, on an average each farmer devotes 8 percent of cropped land for cultivation of rajmash resulting in 1.65 qtls. of this crop. Of the area under this crop about 71.73 percent is irrigated and like previous crops, rajmash in Pooh and Kalpa is grown entirely under irrigated conditions. Only 14.04 percent of area under rajmash is irrigated in Nichar block. Table 4.9 reveals that in Pooh at overall level 7 percent of land is put under this crop by each farmers which results in production of 1.37 qtls. of rajmash. In Kalpa average area of 8 percent under this crop is yielding 1.59 qtls. of produce and in Nichar nearly 9 percent produces 1.98 qtls. of final output. The productivity of this crop is highest in Nichar (1.03 qtls./bigha) followed by Pooh (0.87 qtls./bigha) and Kalpa (0.82 qtls./bigha) respectively. Again the role of irrigation can not influence the productivity of this crop.

Wheat

Each farmer in Kinnaur averaged 16 percent of his total cropped area under wheat resulting an average production of 3.47 qtls. About 73.40 percent of area under wheat is irrigated at overall level. However, cent percent area under wheat is irrigated in Pooh and Kalpa but this percentage is 19.13 in Nichar block. The area allocation for wheat is seen to increase with holding size at overall level of Kalpa block. In Pooh 15 percent of land results in 3.18 qtls. of wheat and the farmers of Kalpa block put an average of 15 percent of land for cultivation of wheat and are able to produce 3.72 qtls. These figures for Nichar are 17 percent and 3.45 qtls. The productivity of wheat in Kalpa block is 0.95 qtls./bigha and in Pooh and Nichar blocks it is 0.91 qtls./bigha each.

Barley

Like the trend of other crops, barley is also grown under entirely irrigated conditions in Pooh and Kalpa blocks. Only 14.92 percent of land under barley is irrigated in Nichar block. In Pooh on an average 14 percent of land yields 2.99 qtls. of barley and in Kalpa nearly 12 percent of land put under this crop results in 3.26 qtls. of crop. These figures for Nichar are observed to be 3.47 bighas and 3.52 Qtls. At district level average land utilised for cultivation of barley is 14 percent and output from this allocation is 3.26 qtls. Again land put under barley increases with holding size. The productivity of barley is highest in Kalpa block (1.01 qtls./bigha). followed by Nichar (1.01 qtls./bigha) and Pooh (0.93 qtls./bigha).

The crops discussed earlier are those which are grown mainly for self consumption. But there are many other crops which have commercial value and are produced mainly for the market. Thesc crops are presented in the following text and in Table 4.10 for analytical details.

Peas

The average area cultivated under peas crop is 8 percent and 7 percent in Pooh and Kalpa blocks respectively. The area under peas in these blocks is fully irrigated. Each farmer devotes 5 percent of holding to peas crop and total average irrigated area is 67.22 percent in the district. The production of peas crop is negligible in Nichar block.

Potato

The farmers of Kinnaur devote 4 percent land for potato crop and produce 8.7 qtls. showing a productivity of 5 qtls. per bigha. The area under potato is 8 percent, 7 percent and 7 percent and productivity is 5 qtls./bigha, 4.99 qtls./bigha and 4.98 qtls./bigha in Pooh, Kalpa and Nichar blocks respectively.

Vegetables

It is observed that people of Kinnaur district have not shown much interest in the vegetable production except potatoes and peas. They use to produce minor vegetables for domestic consumption only. So we have included this system in the agriculture sector. The climatic condition is most suitable for vegetables like cabbage, cauliflower, tomato etc.

Fruit Crops

Traditionally, district Kinnaur is well known for best quality of fruits. The production and value of various fruits has been dealt as under.

Area under Horticulture

The area under horticulture of the sampled farmers and the number of fruit plants have been presented in Table 4.11. It may be seen from the table that at overall level each farmer has 37 percent of land under fruit trees. The area under fruit in individual block is 35 percent, 37 percent and 38 percent in Pooh, Kalpa and Nichar respectively. The total number of fruit trees at overall level are observed to be 241.15/farm of which 226 are apple trees and 15.39/farm are other fruit trees. Similarly in Pooh block total number of fruit trees are 253.10/farm of which 231.84/farm are apple trees. In Kalpa each farmer, on an average, has 240.29 apple and 9.39/farm other fruit trees. In Nichar average number of apple trees are 203.90/farm and 16.73/farm other fruit trees. About 75 percent of apple trees are in bearing stage in all the blocks.

Fruit Production

The apple production dominates the fruit production scene in district Kinnaur as well as in each block. Table 4.12 depicts that average production of apple at small farm level of district Kinnaur is 192.50 boxes per farm whereas this figure is 192.90, 220.37 and 166.36 boxes in Pooh, Kalpa and Nichar blocks respectively. The production of other fruits is comparatively very small. The next important fruit production is grapes at 45.55 kg. and almond at 19.59 kg. per farm. The production of other fruits at overall level and at block level for different categories of farmers have also been presented in this table.

Major and Minor Implements

This section deals with adoption of technological practices in the farming system (Tables 4.13 and 4.14). Although people of Kinnaur generally use the traditional implements, the government has paid proper attention by providing subsidy and another financial help to purchase modern equipments. It is now generally observed that in Kinnaur district government has encouraged the farmers to use modern equipments like power sprayer, modern scissors and other modern equipments.

Tree Inventory

Trees are the integral part of the hill agriculture and its existence is very crucial in meeting out the fuelwood, fodder and miscellaneous needs of the farm households. Therefore, inventory of fruits and forest trees of the sample households has been discussed in Tables 4.15 and 4.16.

Consumption Pattern

The consumption pattern of sample households has been divided into three sections. The first section is most important and includes cereals whereas the second part pertains to other food articles, and finally the non-food items. The details are presented as follows:

Cereals

Keeping in view the importance of cereals in the consumption pattern, both quantities and their values have been studied and presented in Tables 4.17 and 4.18. In cereal consumption domestic produce as well as market purchases have been included in different tables. It may be seen from the tables that on an average each sample household consumes 8.22 qtls. of wheat per annum/farm valued at Rs. 4868.56 and per household consumption of wheat is seen to increase with the holding size. This is also true for individual block where total consumption of wheat is observed to be 7.36 qtls., 8.77 qtls. and 8.4 qtls. per farm/year in Pooh, Kalpa and Nichar blocks respectively. The consumption of rice is almost same at overall level i.e. 7.17 qtls. per household per year and valued at Rs. 7923.53. The consumption at block level is also almost similar, 7.22 qtls. in Pooh, 7.18 qtls. in Kalpa and 7.12 qtls. per year/household in Nichar. Due to almost same quantities, the value is almost similar. Like wheat the consumption of rice is also seen increasing with holding size. Maize is the next cereal under consideration. In district Kinnaur each household consumes 1.21 qtls. of maize per year and it is used totally from domestic products with the value of Rs. 725.73. The highest consumption of maize is in Nichar where it is 1.66 qtls. valued at Rs. 999.21 followed by Kalpa 1.15 qtls. valued at Rs. 691.61 and Pooh 0.77 qtls. valued at Rs. 469.76. The consumption of barley is found to be 3.26 qtls. per household/year at district level and valued at Rs. 1631.42. At block level the highest consumption of this cereal is in Nichar block.

Pulses

The average consumption of pulses like mash, moong & rongi, lentil and gram in this district is 15.87 kg., 3.47 kg., 13.14 and 8.80 kg. worth Rs. 321.76, 69.34, 285.09 and 168.01 respectively. The major part of the pulses consumed is rajmash valued at Rs. 1230.57 per farm/year. Alongwith pulses, potatoes & peas and other vegetables of Rs. 521 and 429.73 per annum respectively are also consumed (Tables 4.19 and 4.20).

Miscellaneous Food Items

In the next group are considered other food articles. The details have been presented in Table 4.18 which depicts that per year expenditure of each household for sugar at overall level is Rs. 790.02. The highest expenditure on sugar is in Nichar (Rs. 907) followed by Pooh (Rs. 898.97) and Kalpa (Rs. 596.68). The expenditure on edible oil is Rs. 2120.64 per year/household in the district level. At block level it ranges between Rs. 1975.00 in Kalpa, Rs. 2264.00 in Nichar and Rs. 2143 in Pooh. Table 4.21 reveals that the sample farmers at overall level consumes milk worth Rs. 106.33 per year and consumption of ghee is of about Rs. 208 per year. Each household spends Rs. 63 on eggs and Rs. 318.02 per year on vegetables. Fruits of worth Rs. 128 are also consumed. The expenditure on spices is little high at Rs. 697.47 per year per household. Tea/coffee consumes Rs. 499.77 of income of each household every year. Rs. 601.53 are spent on tobacco and detergents by each household per year. Meat worth Rs. 2482 is consumed by each household every year and expenditure on liquor is Rs. 803. The category-wise and block-wise details of consumption on these items may also be referred to from Table 4.22.

Non-Food Items

The final group of consumption is that of non-food items like clothing, education and utilities etc. and the details have been presented in Table 4.23. The table reveals that, at overall level, each household has to spend Rs. 720.21 on repair of their house which may have necessitated by snowfall. In addition they have to spent Rs. 157.59 per year on the repair of implements necessary for agriculture. Clothing and footwears are observed to be one of the main item of expenditure. Each household spends Rs. 7332 per year on clothing and footwears of the family members. But the highest expenditure is on education of children.

This head consumes Rs. 12018 from every household per year Rs. 842.66 are spend every year on paying electricity and telephone bills. Ritual ceremonies play an important role in the cultural lives of the inhabitants of Kinnaur, consequently they spent Rs. 5949 each year on this account. Finally, for keeping good health they have to spent on medicines etc. The table also presents the details of different categories in different blocks and may be referred to from this. Per household expenditure on insecticide and pesticide, tree spray oil and fertilizer is, on an average, Rs. 1580; concentrate fed to animals of Rs. 784 and other veterinary expenditure of Rs. 919 (Table 4.24).

Livestock Profile

Animal husbandry is an integral part of farms in the hills. Every household in district Kinnaur is seen to have some type of livestock and average number of livestock owned by the sampled farmers have been presented in Table 4.25. It may be seen from the table that each household, on an average, owns 35.24 heads of livestock. Among this, sheep are in highest number (16.37) followed by goats (10.16). In addition, each household has 1.18 cows in milk and 1.24 dry cows plus the young stock. Young stock includes the animals less than 3 years. There are 1.30 bullocks on each farm. The highest number of livestock is seen in Pooh block (52.82) followed by Nichar (37.17) and Kalpa block (19.11). No definite pattern can be established between size of holding and number of livestock in any of the block.

Migration

Poor and landless people, who generally own flocks of sheep and goats are unable to provide sufficient feed other than grazing in open pastures. For the extraction of maximum potential capacity out of the sheep and goats, it requires maintenance on rich and nutritious pastures, that are not easily available in Kinnaur. Due to the lack of these open pastures, the flock masters move with their flocks to high hills during summer and towards low hills during winters. In Himachal Pradesh, they move to Poanta,

Kinnaur
Routes of migration of Sheep and Goats

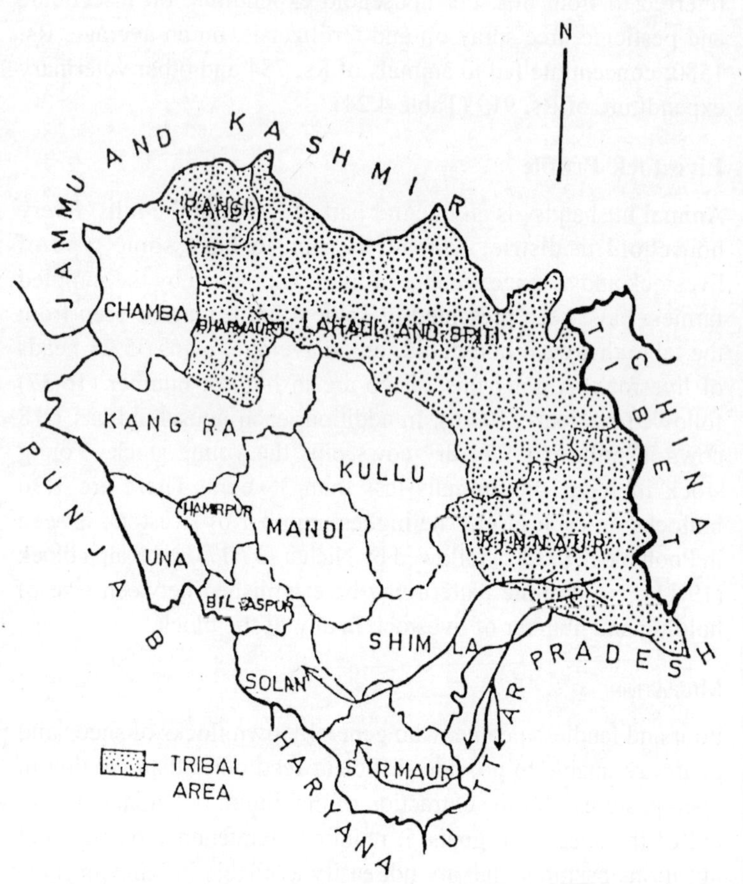

Sataun and Sarahan in Sirmaur and to Bilaspur, Mandi and Solan district. Earlier Dehradun was also their destination for migration but now Uttaranchal Government has stopped to issue entry permits. The migrants move in groups consisting of 5-20 flocks and generally male members of the family accompany the flocks during migration.

Migration generally starts in the end of October. Their journey periods depend on the nature of routes followed and the problems encountered by them. They have to get permits to use the forest lands for grazing. The main reason for the migration to district Sirmaur and Solan is due to the better quality of grasses in the pastures.

The animals are usually grazed for 7-8 hours but during winters this grazing period decreases resulting in weak health of the flocks. The flock is fed with salt and mustard oil, usually once in a week, and each ram or buck is fed with cereals (wheat, maize, barley, ogla, bharat, phapra etc.) everyday to maintain their health. The grounded salt is spread on the rocks for the animals to lick. Reason behind salt feeding is not clear to them and it may be due to the deficiency of iodine and minerals in the grasses of the area. After salt-feeding, the animals are not allowed to drink water. By this practice the animals eat more green grass and remain in better health. The migrants spend 3-4 months in lower hills and start their journey back to their native place at the end of April. Same route is generally followed for back journey.

During summer the grazing place is in alpine region at an altitude of 3050-4570 metres above msl near Indo-Tibet border. In this area grasses are of good quality and are in abundance. Except some salt, once in a week, nothing extra is given to the animals, because the grasses of high hills are considered to contain all required mineral elements.

Household Activities

The various household activities carried by males, females and children have been presented in Tables 4.26 to 4.29. It is seen from the tables that maximum household activities are generally contributed by females. Table 4.26 depicts that 94 percent females, 58 percent males and 13 percent children are devoting time for the child care. In washing of clothes 96 percent females, 35 percent males and 11 percent children are involved at overall level in the study area.

In purchasing of household items, male category is involved maximum (93%) and other household activities like rope-making and spinning etc. are generally done by all the family members.

Cooking is predominantly done by females and helped by males and children. Similar situation is seen to prevail for fetching of water and collection of fuelwood. In cooking, fetching of water and collection of wood 92 percent women are involved.

The labour used for collection of green fodder, dry fodder and green tree leaves to be used for fodder has been presented in Table 4.28. This table depicts that collection of green fodder is predominantly done by females in each block. At overall level, in collection of green fodder from own field 86 percent females and 74 percent of males are involved. In the collection of dry fodder from forest 79 percent females and 71 males are involved. The females are also predominately used for collection of green leaves from trees. Nearly 82 percent of females and 68 percent males collect the green leaves from own fields and forest respectively. This clearly indicates that the burden of fodder collection lies heavily on the shoulder of females. Same situation is also observed in individual blocks. Other household activities like grazing of livestock and milking, all family members are contributing. Open grazing practice is generally prevalent in the study area and 89 percent males are devoting their services, at overall level, in milking 94 percent females complete this activity (Table 4.29).

It may, therefore, be concluded that burden of household activities is generally carried by females but are normally assisted by males and children. The blockwise and farm category-wise details have also been presented in different tables.

Table 4.1. Selected social characteristics of sample households

(%)

Block/ farm size	Average person per family (No.)	Religion		Caste		Family type		Marital status	
		H	B and others	S.C.	Gen. (S.T.)	Nuc.	Joint	M	UM
Pooh Block	**6.15**	**61.77**	**38.23**	**27.93**	**72.07**	**79.41**	**20.59**	**92.64**	**7.36**
Marginal	4.87	81.48	18.52	25.92	74.08	85.19	14.81	92.59	7.41
Small	6.08	48.58	51.42	31.42	68.58	82.85	17.15	91.42	8.58
Medium	12.34	50	50	16.60	83.40	33.30	66.70	100	–
Kalpa Block	**6.73**	**87.95**	**12.05**	**38.54**	**61.46**	**74.27**	**25.73**	**96.39**	**3.61**
Marginal	5.64	93.54	6.46	32.25	67.75	96.78	3.22	95.78	3.22
Small	6.06	81.09	18.91	45.94	54.06	72	28	94.59	5.41
Medium	10.62	93.3	6.7	33.3	66.70	33.34	66.66	100	–
Nichar Block	**7.07**	**89.19**	**10.81**	**28.38**	**71.62**	**68.92**	**31.08**	**97.30**	**2.70**
Marginal	5.14	95.65	4.35	39.14	60.86	91.31	8.69	100	–
Small	7.76	87.5	12.5	25	75	60	40	95	5
Medium	8.61	81.82	18.18	18.18	81.82	54.55	45.45	100	–
All Blocks	**6.67**	**80.44**	**19.56**	**31.99**	**68.01**	**74.06**	**25.94**	**95.55**	**4.45**
Marginal	5.24	90.12	9.88	32.10	67.90	91.36	8.64	96.30	3.70
Small	6.67	73.22	26.78	33.92	66.08	71.10	28.90	93.75	6.25
Medium	10.25	81.24	18.77	24.97	75.03	40.62	59.38	100.00	0.00

Abbreviations used: H= Hindu, B= Buddhist, S.C. = Scheduled Castes, S.T.= Scheduled Tribes, Nuc.= Nucleous, M= Married, UM = Un-Married.

Table 4.2. Male educational level of sample households

(%)

Block/farm size	Educational level						Total	
	R-W	P	M	G	P.G.	Total (No.)	Literate	Illiterate
Pooh Block	**12.11**	**25.84**	**30.36**	**16.81**	**5.37**	**86.53**	**90.49**	**9.51**
Marginal	5.79	23.19	31.89	18.84	5.79	69	85.5	14.5
Small	16.83	28.97	29.91	14.02	3.73	107	93.46	6.54
Medium	13.05	19.56	26.08	23.92	13.04	46	95.66	4.34
Kalpa Block	**17.62**	**16.39**	**33.09**	**18.63**	**1.99**	**89.35**	**87.70**	**12.30**
Marginal	16.46	20.25	35.44	12.66	–	79	84.81	15.19
Small	19.8	12.87	32.67	23.76	0.99	101	90.09	9.91
Medium	14.63	17.07	29.26	18.29	8.55	82	87.8	12.2
Nichar Block	**16.08**	**26.67**	**28.30**	**19.18**	**3.33**	**105.00**	**93.56**	**6.44**
Marginal	15	21.67	40	10	5	60	91.67	8.33
Small	15.07	30.82	20.55	25.34	2.74	146	94.52	5.48
Medium	22	22	32	16	2	50	94	6
All Blocks	**15.45**	**22.63**	**30.69**	**18.26**	**3.45**	**93.64**	**90.47**	**9.53**
Marginal	12.49	21.63	35.55	13.96	3.35	70.27	86.99	13.01
Small	17.18	24.31	27.48	21.28	2.47	118.95	92.73	7.27
Medium	16.87	19.23	29.61	18.56	7.14	64.25	91.41	8.60

Abbreviations used: R-W = Read & Write, P = Primary, M = Matric, G = Graduate, P.G. = Post Graduate.

Table 4.3. Female educational level of sample households

(%)

Block/farm size	Educational level						Total	
	R-W	P	M	G	P.G.	Total (No.)	Literate	Illiterate
Pooh Block	**18.95**	**24.60**	**29.19**	**10.17**	**0.53**	**84.50**	**83.45**	**16.55**
Marginal	15.87	23.8	23.8	12.69	–	63	76.16	23.84
Small	21.49	24.3	34.58	6.55	–	107	86.92	13.08
Medium	18	30	22	20	6	50	96	4
Kalpa Block	**18.52**	**24.71**	**26.64**	**10.85**	**3.37**	**97.45**	**84.09**	**15.91**
Marginal	5	30	26.25	7.5	3.75	80	72.5	27.5
Small	26.61	20.16	28.23	12.1	3.22	124	90.32	9.68
Medium	26.48	25	23.52	14.71	2.94	68	92.65	7.35
Nichar Block	**24.36**	**26.26**	**26.25**	**4.04**	**1.87**	**102.01**	**82.78**	**17.22**
Marginal	14.93	28.36	31.34	4.48	1.49	67	80.6	19.4
Small	28.46	24.82	24.82	1.46	1.46	137	81.02	18.98
Medium	29.17	27.08	20.83	12.5	4.17	48	93.75	6.25
All Blocks	**20.57**	**25.19**	**27.28**	**8.41**	**2.02**	**95.04**	**83.46**	**16.54**
Marginal	11.44	27.47	26.88	8.37	1.86	70.64	76.02	23.98
Small	25.67	23.12	29.00	6.57	1.59	123.33	85.94	14.06
Medium	25.81	26.65	22.31	14.94	3.94	57.75	93.66	6.34

Abbreviations used: R-W = Read & Write, P = Primary, M = Matric, G = Graduate, P.G. = Post Graduate.

Table 4.4. Overall educational level of sample households

Block/farm size	Total education (No.)			Total (%)	
	Male	Female	Total	Literate	Illiterate
Pooh Block	**86.53**	**84.50**	**171.03**	**87.06**	**12.94**
Marginal	69	63	132	81.06	18.94
Small	107	107	214	90.19	9.81
Medium	46	50	96	95.83	4.17
Kalpa Block	**89.35**	**97.45**	**186.80**	**85.85**	**14.15**
Marginal	79	80	159	78.62	21.38
Small	101	124	225	90.22	9.78
Medium	82	68	150	90	10
Nichar Block	**105.00**	**102.01**	**207.01**	**88.19**	**11.81**
Marginal	60	67	127	85.83	14.17
Small	146	137	283	87.99	12.01
Medium	50	48	98	93.88	6.12
All Blocks	**93.64**	**95.04**	**188.68**	**86.99**	**13.01**
Marginal	70.27	70.64	140.91	81.48	18.52
Small	118.95	123.33	242.28	89.41	10.59
Medium	64.25	57.75	122.00	92.43	7.57

Table 4.5. Primary occupation of sample households

(%)

Block/farm size	Occupation					
	Agriculture	Horticulture	Livestock	Service	Trade	Non-farm labour
Pooh Block	**35.29**	**55.88**	**1.48**	**5.88**	**0.00**	**1.47**
Marginal	37.03	55.55	3.72	3.7	–	–
Small	34.28	54.30	–	8.57	–	2.85
Medium	33.3	66.60	–	–	–	–
Kalpa Block	**34.88**	**59.10**	**1.20**	**2.42**	**2.41**	**0.00**
Marginal	41.93	48.38	3.22	3.25	3.22	–
Small	35.00	59.6	–	2.7	2.7	–
Medium	20	80	–	–	–	–
Nichar Block	**50.05**	**39.14**	**2.70**	**8.11**	**0.00**	**0.00**
Marginal	60.86	39.14	–	–	–	–
Small	47.5	32.5	5	15	–	–
Medium	36.7	63.3	–	–	–	–
All Blocks	**39.99**	**51.56**	**1.78**	**5.34**	**0.89**	**0.44**
Marginal	45.67	48.15	2.47	2.48	1.23	0.00
Small	39.24	48.27	1.79	8.93	0.89	0.89
Medium	28.23	71.75	0.00	0.00	0.00	0.00

Table 4.6. Secondary occupation of sample households

(%)

Block/farm size	Occupation					
	Agriculture	Horticulture	Livestock	Service	Trade	Non-farm labour
Pooh Block	**51.41**	**22.04**	**16.19**	**8.86**	**1.46**	–
Marginal	55.50	29.62	7.4	7.48	–	–
Small	54.23	14.27	22.9	8.6	–	–
Medium	16.6	33.2	16.6	16.6	16.6	–
Kalpa Block	**50.59**	**19.28**	**16.88**	**9.64**	**3.61**	–
Marginal	51.61	12.9	25.8	6.45	3.23	–
Small	43.24	27.04	13.51	10.81	5.4	–
Medium	66.6	13.33	6.74	13.33	–	–
Nichar Block	**29.72**	**25.44**	**20.27**	**16.34**	**8.23**	–
Marginal	17.39	39.13	17.39	17.40	8.69	–
Small	27.5	20	27.5	17.5	7.5	–
Medium	63.6	16.6	–	9.9	9.9	–
All Blocks	**43.98**	**22.14**	**17.79**	**11.61**	**4.48**	–
Marginal	43.19	25.92	17.28	9.90	3.70	–
Small	41.05	20.54	21.44	12.51	4.46	–
Medium	56.19	18.18	6.27	12.76	6.52	–

Table 4.7. Land utilisation pattern of sample households

(Bigha/%)

Block/farm size	Land holding			Land utilisation (Bigha)					
	Total land (Bigha)	Irrigated (%)	Un-irrigated (%)	Culti-vated	Un-cultivated	Orchard	Under forest	Pasture	Ghasni
Pooh Block	**19.89**	**86.17**	**13.83**	**17.03**	**2.87**	**7.95**	-	**2.57**	**0.58**
Marginal	8.81	91.17	8.83	8.29	0.52	4.14	-	0.29	0.96
Small	20.08	82.64	17.36	17.54	2.54	8.77	-	2.42	0.11
Medium	68.66	84.22	15.78	53.33	15.33	20.33	-	13.66	1.66
Kalpa Block	**21.86**	**81.96**	**18.04**	**17.88**	**3.98**	**9.52**	**0.25**	**3.23**	**0.49**
Marginal	9.96	78.64	21.36	8.54	1.42	4.51	-	0.48	0.94
Small	19.83	81.74	18.24	16.48	3.35	9.94	0.24	3.05	.054
Medium	51.46	89.37	10.63	40.66	10.8	18.86	0.8	9.33	0.66
Nichar Block	**19.46**	**11.44**	**88.56**	**17.75**	**1.70**	**8.57**	**0.68**	**0.59**	**0.42**
Marginal	9.91	5.26	94.74	9.04	0.87	5.17	0.43	0.173	0.26
Small	19.75	16.83	83.17	18.37	1.37	8.5	0.62	0.50	0.25
Medium	38.36	4.73	95.27	33.72	4.64	15.90	1.45	1.80	1.36
All Blocks	**20.48**	**60.04**	**39.96**	**17.58**	**2.89**	**8.73**	**0.32**	**2.16**	**0.50**
Marginal	9.56	61.98	38.02	8.60	0.96	4.57	0.12	0.33	0.75
Small	19.88	58.84	41.16	17.49	2.39	9.06	0.3	1.94	0.14
Medium	50.18	65.81	40.69	40.65	9.53	18.12	0.87	7.55	1.09

Table 4.8. Cropping pattern and production of different crops of sample households

(Bigha/%/Qtls.)

Block/farm size	Maize				Miscellaneous crops			
	Area (Bigha)	Irrigated (%)	Un-irrigated (%)	Total production (Qtls.)	Area (Bigha)	Irrigated (%)	Un-irrigated (%)	Total production (Qtls.)
Pooh Block	**1.83**	**100.00**	**0.00**	**0.77**	**1.01**	**100.00**	**0.00**	**1.02**
Marginal	0.97	100	–	0.44	0.55	100	–	0.99
Small	2.3	100	–	0.93	1.01	100	–	0.86
Medium	2.95	100	–	1.32	3.07	100	–	1.89
Kalpa Block	**1.92**	**100.00**	**0.00**	**1.10**	**1.31**	**100.00**	**0.00**	**0.77**
Marginal	1.2	100	–	0.64	0.45	100	–	0.52
Small	2.1	100	–	1.31	1.02	100	–	0.60
Medium	2.98	100	–	1.56	3.80	100	–	1.69
Nichar Block	**1.88**	**19.37**	**80.66**	**1.66**	**1.31**	**24.03**	**75.97**	**0.87**
Marginal	0.98	16.66	83.44	0.90	0.60	16.67	83.33	0.61
Small	2.1	18.54	81.46	1.82	0.90	22.5	77.5	0.60
Medium	2.97	28.08	71.92	2.70	4.3	45	55	2.39
All Blocks	**1.88**	**73.48**	**26.52**	**1.19**	**1.22**	**75.02**	**24.98**	**0.85**
Marginal	1.06	76.34	23.69	0.65	0.53	76.34	23.66	0.70
Small	2.16	70.91	29.09	1.37	0.97	72.32	27.68	0.68
Medium	2.97	75.28	24.72	1.91	3.84	81.09	18.91	1.78

Table 4.9. Cropping pattern and production of different crops of sample households

(Bigha/%/Qtls.)

Block/farm size	Rajmash				Wheat				Barley			
	Area (Bigha)	Irri-gated (%)	Un-irri-gated (%)	Total produc-tion (Qtls.)	Area (Bigha)	Irri-gated (%)	Un-irri-gated (%)	Total produc-tion (Qtls.)	Area (Bigha)	Irri-gated (%)	Un-irri-gated (%)	Total produc-tion (Qtls.)
Pooh Block	**1.56**	**100.00**	**0.00**	**1.37**	**3.49**	**100.00**	**0.00**	**3.18**	**3.21**	**100.00**	**0.00**	**2.99**
Marginal	0.86	100	–	0.75	1.93	100	–	1.80	1.2	100	–	0.77
Small	1.70	100	–	1.53	3.93	100	–	3.60	3.95	100	–	4.10
Medium	3.91	100	–	3.18	7.95	100	–	6.90	7.90	100	–	6.51
Kalpa Block	**1.94**	**100.00**	**0.00**	**1.59**	**3.99**	**100.00**	**0.00**	**3.72**	**3.20**	**100.00**	**0.00**	**3.26**
Marginal	0.98	100	–	0.85	1.98	100	–	2.10	0.76	100	–	0.90
Small	1.95	100	–	1.55	4.02	100	–	3.80	3.72	100	–	3.90
Medium	3.92	100	–	3.2	7.62	100	–	6.88	6.97	100	–	6.55
Nichar Block	**1.91**	**14.04**	**85.96**	**1.98**	**3.77**	**19.13**	**80.87**	**3.45**	**3.47**	**14.92**	**85.08**	**3.52**
Marginal	.90	18.36	81.64	.93	2.1	32.30	67.70	1.90	0.94	17.31	82.69	1.02
Small	1.93	10.84	89.16	2.1	3.85	12.90	87.10	3.50	3.80	12.15	87.85	4.03
Medium	3.98	16.66	83.34	3.75	6.97	14.24	85.76	6.50	7.53	20	80	6.88
All Samples	**1.82**	**71.73**	**28.27**	**1.65**	**3.74**	**73.40**	**26.60**	**3.47**	**3.29**	**72.02**	**27.98**	**3.26**
Marginal	0.92	76.82	23.18	0.84	2.00	80.78	19.22	1.94	0.96	76.52	23.48	0.89
Small	1.86	68.16	31.84	1.74	3.93	68.89	31.10	3.63	3.82	68.63	31.38	4.01
Medium	3.94	71.35	28.65	3.39	7.46	70.52	29.48	6.75	7.34	72.50	27.50	6.66

Table 4.10. Cropping pattern and production of different crops of sample households

(Bigha/%/Qtls.)

Block/farm size	Potato				Peas				Other vegetables			
	Area (Bigha)	Irri-gated (%)	Un-irri-gated (%)	Total produc-tion (Qtls.)	Area (Bigha)	Irri-gated (%)	Un-irri-gated (%)	Total produc-tion (Qtls.)	Area (Bigha)	Irri-gated (%)	Un-irri-gated (%)	Total produc-tion (Qtls.)
Pooh Block	**1.80**	**100.00**	**0.00**	**9.02**	**1.75**	**100.00**	**0.00**	**2.15**	**0.13**	**100.00**	**0.00**	**47.48**
Marginal	1	100	–	5.0	1	100	–	1	.08	100	–	30
Small	2.8	100	–	10.5	1.9	100	–	2.3	.15	100	–	56.25
Medium	3.70	100	–	18.5	4.2	100	–	6.5	.20	100	–	75
Kalpa Block	**1.89**	**100.00**	**0.00**	**9.44**	**1.87**	**100.00**	**0.00**	**2.49**	**0.15**	**100.00**	**0.00**	**56.20**
Marginal	0.90	100	–	4.5	0.90	100	–	1.2	0.09	100	–	33.75
Small	1.90	100	–	9.5	1.93	100	–	2.35	0.20	100	–	75
Medium	3.90	100	–	19.5	3.70	100	–	5.50	0.15	100	–	56.25
Nichar Block	**1.52**	**9.73**	**90.27**	**60.58**	**0.00**	**0.00**	**0.00**	**0.00**	**0.15**	**100.00**	**0.00**	**56.66**
Marginal	0.75	8.33	91.67	3.75	-	-	-	-	0.12	100	-	45
Small	1.30	9.27	90.73	6.5	-	-	-	-	0.15	100	-	56.25
Medium	3.90	14.35	85.65	19.5	-	-	-	-	0.22	100	-	82.5
All Samples	**1.74**	**70.31**	**29.69**	**8.70**	**1.22**	**67.11**	**0.00**	**1.57**	**0.14**	**67.11**	**0.00**	**53.72**
Marginal	0.89	73.97	26.03	4.45	4.68	71.60	-	0.79	0.10	71.60	-	35.69
Small	1.75	67.60	32.40	8.74	1.23	64.29	-	1.50	0.17	64.29	-	62.44
Medium	3.86	70.56	29.44	19.31	2.52	65.63	-	3.80	0.18	65.63	-	68.79

Table 4.11. Area and number of fruit plants of sample households

(Average/farm)

Block/farm size	Area under fruits (Bigha)	Apple plants (No.)			Other fruits/ plants (No.)	Total fruits/ plants (No.)
		Bearing	Non-bearing	Total plants		
Pooh Block	**7.95**	**158.67**	**73.17**	**231.84**	**21.26**	**253.10**
Marginal	4.14	80.68	74.51	155.19	13.32	168.51
Small	8.77	175.23	66.91	242.14	13.54	255.28
Medium	20.33	413	103.66	516.66	102	618.66
Kalpa Block	**9.52**	**196.27**	**44.02**	**240.29**	**9.39**	**249.67**
Marginal	4.51	96.62	30.80	127.42	7.74	135.16
Small	9.94	196.8	63.78	260.58	8.1	268.68
Medium	18.86	400.89	22.6	423.49	15.96	439.45
Nichar Block	**8.57**	**146.93**	**56.97**	**203.90**	**16.73**	**220.63**
Marginal	5.17	94.69	29.22	123.91	20.82	144.73
Small	8.5	155.38	55.37	210.75	16	226.75
Medium	15.90	225.45	120.81	346.26	10.8	357.06
All Blocks	**8.73**	**168.68**	**57.09**	**225.77**	**15.39**	**241.15**
Marginal	4.57	90.76	44.92	135.68	13.31	148.99
Small	9.06	175.27	61.75	237.02	12.62	249.64
Medium	18.12	342.85	71.56	414.41	30.32	444.73

Table 4.12. Total fruit products of sample household

(Qty.)

Block/farm size	Apple (boxes)	Almond (kg)	Grapes & Bemmi (kg)	Apricot (kg)	Plum (boxes)	Walnut (kg)	Pears (boxes)
Pooh Block	**169.26**	**37.34**	**64.33**	**3.97**	**0.00**	**5.70**	**5.57**
Marginal	114.70	17.33	75	2.96	-	-	4.41
Small	192.90	44.77	55	3.14	-	8.5	4.57
Medium	276.88	84	70.75	13.3	-	15.0	16.67
Kalpa Block	**206.09**	**10.80**	**49.02**	**1.45**	**0.00**	**8.67**	**3.13**
Marginal	118.96	9.80	51.50	.98	-	5.16	2.58
Small	220.37	5.13	47.25	1.35	-	5.67	2.70
Medium	350.94	26.86	48.25	2.66	-	23.33	5.33
Nichar Block	**154.18**	**13.15**	**24.40**	**0.00**	**21.57**	**15.53**	**3.92**
Marginal	80.70	8.82	29.18	-	21.7	17.39	6.52
Small	166.36	12.75	23.50	-	20.0	15	2.5
Medium	263.55	23.63	17.67	-	27.0	13.6	3.63
All Blocks	**177.89**	**19.59**	**45.55**	**1.73**	**7.09**	**10.03**	**4.13**
Marginal	106.68	12.03	53.00	1.36	6.16	6.91	4.31
Small	192.50	20.24	41.19	1.43	7.14	9.89	3.21
Medium	307.01	36.46	41.96	3.74	9.28	18.42	6.87

Weight of boxes approx.—Apple box = 18 kg, Almond = 7 kg, Plum = 7 kg, Pear = 15 kg.

Table 4.13. Major implements per household of sample households

(Nos./farm)

Block/farm size	Implements									
	Traditional plough	Iron plough	Planker	Yoke	Harrow	Thresher	Foot sprayer	Power sprayer	Scissor	Saw
Pooh Block	**1.59**	**0.82**	**0.01**	**1.38**	**0.72**	**0.26**	**0.87**	**0.14**	**1.14**	**1.22**
Marginal	1.48	0.92	0.03	1.25	0.48	0.10	.77	.03	.92	1.11
Small	1.54	.65	–	1.34	.85	.29	.83	0.14	1.05	0.94
Medium	2.33	1.33	–	2.16	1	.83	1.5	0.67	2.66	3.3
Kalpa Block	**1.52**	**0.73**	–	**1.29**	**0.25**	**0.12**	**0.76**	**0.23**	**1.06**	**0.96**
Marginal	1.35	0.74	–	1.19	0.03	0.03	0.54	0.10	0.95	1.09
Small	1.56	0.67	–	1.24	0.54	–	0.72	0.19	1	1
Medium	1.8	0.86	–	1.6	0	0.60	1.33	0.60	1.46	0.60
Nichar Block	**1.70**	**0.78**	**0.01**	**1.46**	–	**0.18**	**0.86**	**0.20**	**1.16**	**1.17**
Marginal	1.7	1	–	1.69	–	–	0.56	–	1.08	1.29
Small	1.7	0.60	0.025	1.32	–	0.25	0.92	0.25	1.15	1
Medium	1.72	1	–	1.5	–	0.27	1.27	0.45	1.40	1.6
All Blocks	**1.60**	**0.77**	**0.01**	**1.37**	**0.31**	**0.18**	**0.83**	**0.13**	**1.12**	**1.11**
Marginal	1.49	0.87	0.01	1.35	0.17	0.04	0.62	0.05	0.98	1.15
Small	1.60	0.64	0.01	1.30	0.44	0.18	0.83	0.05	1.07	0.98
Medium	1.87	1.00	–	1.67	0.19	0.53	1.34	.56	1.66	1.45

Table 4.14. Minor implements per household of sample households

(Nos./farm)

Block/farm size	Implements								
	Kudali	Space	Drat	Sickle	Pick-axe	Metal storage bin	Wooden storage bin	Basket	Kilta
Pooh Block	**7.63**	**1.79**	**2.75**	**8.52**	**1.54**	**0.94**	**1.51**	**4.49**	**6.71**
Marginal	5.74	2.25	2.51	9.11	1.33	–	1.66	4.25	5.74
Small	8.6	1.42	2.54	7.45	1.54	1.62	1.14	4.62	6.71
Medium	10.5	1.83	5	12.16	2.5	1.16	3	4.83	11.1
Kalpa Block	**7.92**	**1.77**	**2.16**	**6.91**	**1.55**	**0.61**	**1.38**	**3.31**	**6.15**
Marginal	5.54	1.74	2.06	6.25	1.09	1.03	1.32	3.29	5.61
Small	8.78	1.67	1.40	6.10	0.91	0.135	1.18	3.27	5.70
Medium	10.73	2.06	4.26	10.26	4.06	0.93	2	3.46	8.4
Nichar Block	**8.08**	**1.61**	**2.65**	**7.66**	**1.29**	**0.62**	**1.19**	**3.50**	**6.42**
Marginal	7.52	1.43	2.91	7.08	1.43	1.47	0.87	5.13	6.52
Small	7.95	1.25	2.07	7.37	0.67	0.12	1.27	2.72	6.1
Medium	9.72	3.29	4.20	9.90	3.27	0.636	1.54	2.90	7.36
All Blocks	**7.89**	**1.72**	**2.50**	**7.64**	**1.46**	**0.71**	**1.36**	**3.73**	**5.91**
Marginal	6.17	1.82	2.45	7.44	1.27	0.89	1.31	4.13	5.91
Small	8.43	1.44	2.00	6.98	1.02	0.59	1.20	3.50	5.91
Medium	10.34	2.44	4.38	10.49	3.50	0.87	2.03	3.52	5.91

Table 4.15. Inventory of fruit trees and other tree species (age-wise) under different farm households

(Nos./farm)

Block/farm size	Total No. of trees>5 years				Total No. of trees<5 years				Total trees
	Fodder trees	Fuel trees	Timber trees	Fruit trees	Fodder trees	Fuel trees	Timber trees	Fruit trees	
Pooh Block	**4.05**	**5.69**	**5.48**	**196.63**	**3.73**	**5.55**	**4.45**	**56.26**	**281.85**
Marginal	4.76	6.55	5.03	158.17	2.85	5	3.22	10.34	195.92
Small	4.2	3.94	3.57	180	3	4.62	5.34	75.28	279.95
Medium	–	12	18.66	466.66	12	13.5	4.83	152.00	679.65
Kalpa Block	**4.81**	**6.55**	**3.62**	**198.01**	**5.43**	**5.86**	**3.91**	**51.67**	**279.85**
Marginal	2.41	3.64	3.58	87.25	2.83	2.64	3	47.91	153.24
Small	4.64	7.45	2.24	222.43	6.56	7.1	3.05	46.25	299.72
Medium	10.2	10.33	7.13	366.66	8	9.46	7.93	72.79	492.5
Nichar Block	**6.38**	**6.62**	**3.50**	**178.51**	**5.31**	**7.06**	**4.43**	**41.97**	**252.99**
Marginal	4.65	5	4.43	105.86	4.47	4.86	7.08	38.37	172.22
Small	4.6	6.4	2.03	188.12	4.5	7.45	1.77	38.63	253.50
Medium	16.45	10.81	6.9	295.45	10	10.27	8.54	61.61	420.03
All Blocks	**5.10**	**6.31**	**4.14**	**191.18**	**4.88**	**6.16**	**4.25**	**49.87**	**271.62**
Marginal	3.83	5.00	4.30	116.17	3.30	4.06	4.23	32.68	172.86
Small	4.49	5.98	2.58	196.92	4.71	6.45	3.31	52.60	277.03
Medium	10.44	10.81	9.21	360.93	9.44	10.50	7.56	83.80	502.68

Table 4.16. Inventory of forest plants in own land of sample households

(Average/farm/Nos.)

Block/farm size	Forest plants						
	Ban	Moru	Kharsho	Chil	Deodar	Kail	Thuni
Pooh Block	–	–	**0.14**	**0.76**	**0.32**	**0.23**	–
Marginal	–	–	–	0.18	0.33	0.03	–
Small	–	–	0.28	0.23	0.2	0.31	–
Medium	–	–	–	1.16	1	0.66	–
Kalpa Block	**0.09**	**0.1**	**0.72**	**0.24**	**0.27**	**0.31**	–
Marginal	0.03	0.03	0.03	0.09	0.03	–	–
Small	0.16	0.18	0.24	0.27	0.37	0.43	–
Medium	0.06	0.06	3.33	0.47	0.53	0.66	–
Nichar Block	**0.31**	**0.33**	**0.16**	**0.13**	**0.18**	**0.24**	**0.18**
Marginal	0.39	0.09	0.17	0.13	0.28	0.26	–
Small	0.1	0.3	0.05	0.1	0.13	0.18	0.18
Medium	0.9	1	0.54	0.27	0.36	0.45	0.64
All Blocks	**0.13**	**0.15**	**0.32**	**0.96**	**0.22**	**0.26**	**0.06**
Marginal	0.12	0.04	0.06	0.14	0.19	0.09	–
Small	0.09	0.17	0.18	0.19	0.23	0.3	0.06
Medium	0.34	0.37	1.75	0.78	0.56	0.59	0.22

Table 4.17. Consumption pattern (domestic products) of sample households

(Qtls./Rs./Farm/Year)

Block/farm size	Wheat		Barley		Maize		Total
	Qty.	Value	Qty.	Value	Qty.	Value	value
Pooh Block	**3.18**	**2064.71**	**2.99**	**1495.22**	**0.77**	**469.76**	**4029.69**
Marginal	1.8	1170	0.77	385	0.44	284	1839
Small	3.6	2340	4.1	2050	0.93	557.83	4947.83
Medium	6.9	4485	6.51	3255	1.32	791.98	8531.98
Kalpa Block	**3.72**	**2419.10**	**3.26**	**1629.22**	**1.15**	**691.61**	**4739.93**
Marginal	2.1	1365	0.9	450	0.77	460.77	2275.77
Small	3.8	2470	3.9	1950	1.31	785.97	5205.97
Medium	6.88	4472	6.55	3275	1.56	935.93	8682.93
Nichar Block	**3.45**	**2241.62**	**3.52**	**1759.05**	**1.66**	**999.21**	**4999.88**
Marginal	1.9	1235	1.02	510	0.9	540.97	2285.97
Small	3.5	2275	4.03	2015	1.82	1091.99	5381.99
Medium	6.5	4225	6.88	3440	2.7	1619.99	9284.99
All Blocks	**3.47**	**2253.62**	**3.26**	**1631.42**	**1.21**	**725.73**	**4610.78**
Marginal	1.94	1263.09	0.89	445.37	0.70	424.62	2133.08
Small	3.63	2359.73	4.01	2004.46	1.37	823.97	5188.17
Medium	6.75	4389.53	6.66	3327.97	1.91	1144.09	8861.59

Table 4.18. Expenditure on various items (purchased from market) of sample households

(Qtls./Rs./Farm/Year)

Block/farm size	Wheat		Rice		Sugar (kg.)		Edile oil (kg.)		Total
	Qty.	Value	Qty.	Value	Qty.	Value	Qty.	Value	value
Pooh Block	**4.18**	**2301.49**	**7.22**	**8105.82**	**56.26**	**898.97**	**51.91**	**2143.36**	**13449.64**
Marginal	4.55	2502.25	5.66	7148	36.5	577	36.3	1507	11734.50
Small	4.07	2238.5	7.8	8091.42	65.14	1048.6	58.57	2442.85	13821.37
Medium	3.21	1765.5	10.83	12500	93.33	1475	83.33	3260	17307.16
Kalpa Block	**5.05**	**2775.91**	**7.18**	**7875.89**	**35.85**	**596.68**	**49.97**	**1974.65**	**13223.13**
Marginal	5.02	2761	5.67	6554.8	36.77	683.37	37.6	1470.96	11470.13
Small	5.17	2843.5	7.6	7986.5	21.24	320	55.2	2208	13824.20
Medium	4.8	2640	9.29	10333.3	70	1100	62.66	2440	16513.3
Nichar Block	**4.95**	**2722.43**	**7.12**	**7809.46**	**52.16**	**906.76**	**67.86**	**2263.52**	**13702.15**
Marginal	6.59	3624.5	7.26	7621.73	46.52	756.52	45.86	1991.3	13994.05
Small	4.4	2420	6.7	7530	49.13	871.25	75.8	2292.5	13113.75
Medium	3.52	1936	8.36	9218.18	75	1350	85	2727.3	15231.75
All Blocks	**4.75**	**2614.94**	**7.17**	**7923.53**	**47.38**	**790.02**	**56.44**	**2120.64**	**13449.13**
Marginal	5.31	2919.94	6.12	7055.49	39.45	668.68	39.51	1630.72	12274.84
Small	4.55	2503.19	7.34	7856.25	44.92	744.56	63.61	2311.57	13415.57
Medium	4.06	2234.03	9.26	10356.23	76.09	1256.25	74.22	2692.51	16539.02

Table 4.19. Expenditure on pulses (from market) of sample households

(Kgs./Rs./Farm/Year)

Block/farm size	Mash		Mong & Rongi etc.		Lentil & mix		Gram		Total
	Qty.	Value	Qty.	Value	Qty.	Value	Qty.	Value	value
Pooh Block	**17.95**	**383.52**	**0.38**	**8.03**	**13.37**	**329.26**	**11.23**	**206.01**	**926.83**
Marginal	14.66	273.51	0.96	20.22	12.55	301.48	12.22	217	812.21
Small	19.14	425.57	–	–	11.43	290.71	10.71	208.57	924.85
Medium	25.83	633.3	–	–	28.33	679.16	9.83	141.66	1454.12
Kalpa Block	**14.00**	**328.61**	**4.76**	**78.13**	**12.37**	**304.58**	**8.71**	**167.34**	**878.66**
Marginal	10.35	255.0	0.32	6.77	10.54	261.77	9	182.1	705.64
Small	13.42	322.16	1.49	35.81	9.8	245	6.33	133.1	736.07
Medium	23.0	496.66	22	330	22.5	540	14	221.3	1587.96
Nichar Block	**16.05**	**257.33**	**4.86**	**115.93**	**13.81**	**222.63**	**6.66**	**133.84**	**788.79**
Marginal	13.43	256.21	4.84	117.61	6.6	66.3	7.39	144.78	683.9
Small	12.6	211.25	2.8	75.0	15.2	181.25	2.45	35	527.5
Medium	23.18	527.25	12.4	260.6	23.8	700	20.45	470.4	1958.25
All Blocks	**15.87**	**321.76**	**3.47**	**69.34**	**13.14**	**285.09**	**8.80**	**168.01**	**863.66**
Marginal	12.66	261.51	1.82	42.73	10.09	219.50	9.62	183.14	734.99
Small	15.99	314.86	1.49	38.62	12.24	236.52	6.31	121.65	720.57
Medium	23.59	498.42	14.58	244.27	24.04	621.09	15.44	292.00	1690.15

Table 4.20. Consumption pattern (domestic products) of sample households

(Kgs./Rs./Farm/Annum)

Block/farm size	Rajmash		Misc. crops		Potato & peas		Vegetables (minor)		Total
	Qty.	Value	Qty.	Value	Qty.	Value	Qty.	Value	value
Pooh Block	**43.59**	**1299.11**	**63.31**	**655.93**	**62.61**	**500.88**	**47.48**	**379.85**	**2835.78**
Marginal	19.41	670	40	357.5	61	488	30	240	1755.50
Small	52.57	1515.71	75	656.5	62.5	500	56.25	450	3122.21
Medium	100	2866.66	100	1995.5	70.5	564	75	600	6026.16
Kalpa Block	**39.18**	**1081.08**	**58.30**	**486.13**	**62.85**	**502.80**	**56.20**	**449.64**	**2703.98**
Marginal	16.12	475.16	0.45	292.5	29	232	33.75	270	1269.66
Small	49.51	1237.83	100	663	87.5	700	75	600	3200.83
Medium	61.38	1946.66	75	450	72	576	56.25	450	4442.66
Nichar Block	**34.75**	**1335.27**	**75.54**	**578.65**	**69.99**	**559.95**	**56.66**	**453.24**	**3201.36**
Marginal	26.95	1141.3	60	390	80	640	45	360	2531.3
Small	31.8	1251.5	75	585	68.5	548	56.25	450	2834.5
Medium	61.81	2045.45	110	950	54.5	436	82.5	660	5936.45
All Blocks	**39.06**	**1230.57**	**65.48**	**567.87**	**65.13**	**521.01**	**53.72**	**429.73**	**2907.39**
Marginal	20.29	729.26	30.54	341.85	54.15	433.19	35.69	285.56	1789.85
Small	44.14	1329.55	83.26	633.11	72.90	583.21	62.44	499.55	3045.43
Medium	68.77	2153.12	91.72	911.66	65.70	525.63	68.79	550.31	5253.06

Table 4.21. Items of expenditure and the values of the food items of sample households

(Rs./Farm/Annum)

Block/farm size	Milk	Ghee	Eggs	Vegetables	Fruits	Medicines	Total value
Pooh Block	**129.96**	**192.64**	**77.21**	**247.89**	**123.96**	**252.97**	**1024.62**
Marginal	102.80	201.85	70.37	228	127	276	822.67
Small	155.45	200	78.57	162.87	108.57	180	885.46
Medium	103.50	108.30	100	833.30	200	575	1920.1
Kalpa Block	**109.15**	**160.35**	**39.55**	**449.98**	**135.54**	**90.14**	**984.71**
Marginal	85	79.03	12.9	124.2	16.12	79.03	396.28
Small	130.70	54.04	27.03	540.5	54.05	38.70	845.02
Medium	105.90	590.66	125.50	900	583.33	240	2545.39
Nichar Block	**81.46**	**276.04**	**76.35**	**234.46**	**122.97**	**131.07**	**922.35**
Marginal	95.80	94.8	21.73	91.3	39.13	56.52	399.28
Small	43.47	257.5	75	251.25	90	140	857.22
Medium	189.65	722.40	195.45	472.72	418.18	254.45	2252.85
All Blocks	**106.33**	**208.16**	**63.03**	**318.02**	**127.90**	**152.81**	**976.26**
Marginal	94.00	124.45	34.56	149.46	59.61	138.29	600.38
Small	107.28	172.32	60.27	319.19	83.93	119.03	862.01
Medium	134.24	545.50	144.76	740.62	454.69	307.78	2327.59

Table 4.22. Expenditure on other items of sample households

(Rs./Farm/Year)

Block/farm size	Spices	Tea/coffee	Tobacco & detergent	Meat	Wine	Total exp.
Pooh Block	**636.85**	**552.18**	**674.97**	**2031.76**	**824.99**	**4720.76**
Marginal	578	416.6	566.6	1191.11	1096.3	3848.61
Small	560	608.57	708.57	2571.42	271.42	4719.98
Medium	1350	833.33	966.66	2666.66	2833.3	8649.95
Kalpa Block	**755.72**	**407.83**	**605.99**	**2695.17**	**920.48**	**5385.19**
Marginal	387.09	117.74	512.9	2290.32	877.41	4185.46
Small	925	500	675.6	3013.5	702.7	5816.8
Medium	1100	780	626.66	2746.66	1546.66	6799.98
Nichar Block	**687.84**	**554.73**	**529.05**	**2655.41**	**651.35**	**5078.38**
Marginal	530.43	545.65	545.65	2086.95	108.69	3817.37
Small	692.5	540	497.5	2962.5	750	5442.5
Medium	1000	627.27	609.1	2727.3	1427.3	6390.97
All Blocks	**697.47**	**499.77**	**601.53**	**2481.60**	**803.11**	**5083.48**
Marginal	491.43	338.87	540.10	1866.17	732.09	3968.66
Small	727.90	548.21	622.30	2857.14	584.82	5340.37
Medium	1112.50	737.50	684.37	2725.01	1746.88	7006.25

Table 4.23. Expenditure on non-food items of sample households

(Rs./Farm/Year)

Block/farm size	House tax	Land revenue	Repair of implement	Repair of house	Clothes & foot-wears	Educa-tion	Telephone & electri-city	Ritual ceremo-nies	Total expendi-ture
Pooh Block	**5.55**	**15.78**	**197.03**	**819.85**	**7905.87**	**13272.04**	**854.41**	**7389.71**	**29430.83**
Marginal	3.87	8.61	116.6	364.81	5244.44	8944.40	507.41	5111.10	20301.24
Small	6.59	14.64	231.42	1011.42	8885.7	14142.85	925.71	8585.74	31804.07
Medium	7	54.66	358.33	1750	14166.66	27666.67	2000	10666.6	56669.92
Kalpa Block	**6.15**	**16.57**	**146.98**	**699.98**	**7813.25**	**10855.38**	**827.70**	**6024.13**	**26390.14**
Marginal	5.06	5.96	32.25	258.06	4661.29	7096.7	377.41	4838.8	17275.53
Small	6.89	13.10	193.24	948.60	8675.67	9837.8	959.45	6270.27	26905.02
Medium	6.56	47.08	270	1000	12200	21133.33	1433.33	7866.66	43956.96
Nichar Block	**6.07**	**15.79**	**133.24**	**651.35**	**6264.88**	**12168.93**	**848.64**	**4540.54**	**24629.43**
Marginal	6.13	8.04	13.04	282.6	4260.86	7586.98	647.80	3500	16305.45
Small	6	16.34	173.75	732.5	6090	12575	917.50	4437.5	24948.59
Medium	6.18	30.01	237.27	1127.27	11091	20272.72	1018.18	7090.90	40873.53
All Blocks	**5.94**	**16.08**	**157.59**	**720.21**	**7332.00**	**12017.76**	**842.66**	**5948.90**	**26730.03**
Marginal	4.97	7.43	54.91	300.61	4741.97	7851.82	497.52	4549.41	18008.65
Small	6.48	14.74	198.21	891.05	7817.85	12160.70	933.92	6339.29	27734.24
Medium	6.51	42.63	275.31	1184.37	12187.53	22062.50	1396.87	8124.98	45280.71

Table 4.24. Expenditure on various non-food items of farms (purchased from market)

(Rs./Farm/Annum)

Block/ farm size	Horticultural medicine and spray oil, fertilizer etc.	Concentrate/cakes fed to animals	Veterinary expenditure	Total expenditure
Pooh Block	**1970.83**	**824.35**	**696.25**	**3491.44**
Marginal	1400.6	640	715	2755.6
Small	2034.3	872	640	3546.3
Medium	4166.66	1376	940	6482.66
Kalpa Block	**1606.02**	**781.49**	**997.83**	**3385.35**
Marginal	1129.03	640	865	2634.03
Small	1481.08	792	1090	3363.08
Medium	2900	1048	1045	4993
Nichar Block	**1192.98**	**750.49**	**1033.85**	**2977.32**
Marginal	328.26	720	1075	2123.26
Small	1343.26	728	970	3041.75
Medium	2454.54	896	1180	4530.54
All Blocks	**1580.43**	**784.25**	**918.53**	**3283.21**
Marginal	992.17	662.72	874.63	2529.52
Small	1604.74	794.14	906.52	3305.40
Medium	2984.37	1057.25	1071.72	5113.34

Table 4.25. Number of livestock per household

(Nos./Farm)

Block/farm size	Cows in milk and dry/youngs		Bullocks yaks (Avg. No.)	Horse, mule pony & donkey (Avg. No.)	Goats		Sheep		Avg. No. livestock per house-hold
	Cows in milk	Cows dry/ youngs			Avg. No. goats	Avg. No. youngs	Avg. No. sheep	Avg. No. youngs	
Pooh Block	**1.32**	**1.25**	**1**	**0.95**	**16.17**	**2.90**	**25.59**	**3.63**	**52.82**
Marginal	1.56	1.44	1.04	1.04	17.26	2.52	27.03	3.07	54.96
Small	1.11	1.09	0.92	0.82	12.4	2.46	20	2.97	41.77
Medium	1.50	1.34	1.34	1.34	33.3	7.17	51.67	10	107.67
Kalpa Block	**1.17**	**0.79**	**1.41**	**0.30**	**5.18**	**1.32**	**7.39**	**1.56**	**19.11**
Marginal	1.19	0.87	1.23	0.35	5.16	1.42	7.68	1.61	19.51
Small	0.97	0.92	1.52	0.16	5	0.95	7.3	1.46	18.67
Medium	1.6	0.33	1.5	0.53	5.66	2	7	1.7	20.4
Nichar Block	**1.05**	**1.73**	**1.46**	**0.34**	**10.21**	**1.67**	**17.97**	**2.73**	**37.17**
Marginal	1.13	2.13	1.56	0.22	10.69	2.65	20.86	3.48	42.74
Small	0.90	1.83	1.35	0.40	10.25	1.0	19.25	2.58	37.55
Medium	1.45	0.55	1.64	0.36	9.09	2.09	7.27	1.73	24.18
All Blocks	**1.18**	**1.24**	**1.30**	**0.51**	**10.16**	**1.91**	**16.37**	**2.57**	**35.24**
Marginal	1.30	1.42	1.26	0.54	10.76	2.14	17.87	2.63	37.92
Small	0.99	1.30	1.27	0.45	9.19	1.44	15.54	2.33	32.51
Medium	1.53	0.60	1.52	0.62	12.02	3.00	15.47	3.27	38.02

Table 4.26. Sex-wise sharing of different activities of sample households

(Multiple response/%)

Block/	Child care			Washing clothes			Purchase of households item			Rope making & spinning		
farm size	Male	Female	Children	Male	Female	Children	Male	Female	Children	Male	Female	Children
Pooh Block	**55.73**	**89.70**	**5.88**	**13.23**	**86.76**	**0.00**	**89.71**	**35.29**	**23.64**	**83.25**	**55.44**	**12.87**
Marginal	44.44	77.77	11.11	11.11	100	-	100	25.92	29.92	81.14	11.11	-
Small	68.27	97.14	2.85	8.57	74.28	-	80	37.14	20	82.0	82.0	25.0
Medium	33.33	100	-	50	100	-	100	66.66	16.66	100	100	-
Kalpa Block	**45.78**	**92.77**	**9.64**	**31.34**	**100.00**	**10.84**	**90.36**	**43.37**	**27.70**	**43.50**	**75.91**	**14.45**
Marginal	32.25	100	-	-	100	-	100	-	32.25	0	83.87	9.67
Small	59.45	83.78	21.62	54.08	100	24.32	78.37	67.56	24.32	67.86	59.48	24.32
Medium	40	100	-	40	100	-	100	73.33	26.66	73.33	100	-
Nichar Block	**72.97**	**100.00**	**22.97**	**59.46**	**100.00**	**20.27**	**98.65**	**83.78**	**25.67**	**83.78**	**68.92**	**28.11**
Marginal	100	100	30.43	73.91	100	30.43	100	73.91	30.43	60.86	78.26	34.78
Small	65	100	25	52.5	100	20	97.5	92.5	20	100	55	12
Medium	45.45	100	-	54.54	100	-	100	72.72	36.36	72.72	100	72.72
All Blocks	**57.73**	**94.22**	**12.89**	**35.11**	**96.00**	**10.67**	**92.89**	**54.22**	**25.81**	**68.76**	**67.43**	**18.46**
Marginal	55.55	92.59	12.34	24.69	100.00	8.64	100.00	29.63	30.96	44.33	58.02	13.58
Small	64.19	93.75	16.96	39.29	91.96	15.18	85.71	66.96	21.43	83.76	64.92	20.13
Medium	40.62	100.00	0.00	46.87	100.00	0.00	100.00	71.87	28.12	78.12	100.00	25.00

Table 4.27. Sex-wise sharing of different activities of sample households

(Multiple response/%)

Block/farm size	Cooking			Fetching water			Fuelwood collection		
	Male	Female	Children	Male	Female	Children	Male	Female	Children
Pooh Block	**22.06**	**89.71**	**7.35**	**60.29**	**86.76**	**16.17**	**83.82**	**83.82**	**5.88**
Marginal	22.22	100	18.51	40.74	85.18	25.92	96.29	66.66	14.81
Small	17.14	80	-	77.14	91.42	8.57	74.28	94.28	-
Medium	50	100	-	50	66.66	16.66	83.33	100	-
Kalpa Block	**29.19**	**95.18**	**8.43**	**27.71**	**90.36**	**16.86**	**43.37**	**95.18**	**9.64**
Marginal	19.35	100	-	-	100	-	-	100	-
Small	27.72	89.18	18.9	51.35	86.48	24.32	70.27	89.18	21.62
Medium	53.13	100	-	26.66	80	33.33	66.66	100	-
Nichar Block	**60.81**	**94.59**	**0.00**	**59.46**	**97.30**	**16.22**	**70.27**	**95.94**	**9.46**
Marginal	60.86	82.60	-	69.56	100	39.13	69.56	86.95	30.43
Small	67.5	100	-	67.5	95	-	70	100	-
Medium	36.36	100	-	9.09	100	27.27	72.72	100	-
All Blocks	**37.43**	**93.33**	**5.33**	**48.00**	**91.55**	**16.44**	**64.44**	**92.00**	**8.44**
Marginal	32.09	95.06	6.17	33.33	95.06	19.75	51.85	85.18	13.58
Small	38.62	90.18	6.24	65.18	91.07	10.17	71.43	94.64	7.14
Medium	46.78	100.00	0.00	25.00	84.37	28.12	71.87	100.00	0.00

Table 4.28. Different fodder collection activities (sex-wise) of sample households

(Multiple response/%)

Block/	Green fodder				Dry fodder				Green tree leaves			
	Own field		Forest		Own field		Forest		Own field		Forest	
farm size	Male	Female	Male	Female	Male	Female	Male	Female	Male	Female	Male	Female
Pooh Block	**74.99**	**88.23**	**75.00**	**79.41**	**73.53**	**85.26**	**69.11**	**76.44**	**70.58**	**94.11**	**75.00**	**61.76**
Marginal	85.18	88.88	88.88	70.37	59.25	88.8	62.96	77.7	62.96	88.88	77.7	44.44
Small	68.57	85.71	68.57	82.85	80	80	74.28	71.42	77.14	97.14	77.14	68.57
Medium	66.6	100	50	100	100	100	66.6	100	66.6	100	50	100
Kalpa Block	**77.11**	**88.31**	**63.84**	**87.95**	**53.00**	**89.15**	**78.31**	**67.47**	**92.77**	**97.59**	**36.14**	**54.22**
Marginal	96.77	90.32	70.96	87.09	41.93	93.54	96.77	35.48	96.77	93.54	64.51	96.77
Small	62.16	81.88	64.86	86.48	62.16	83.78	72.97	81.08	100	100	-	-
Medium	73.33	100	46.6	93.33	53.3	93.33	53.33	100	66.66	100	66.66	100
Nichar Block	**70.27**	**82.57**	**68.92**	**100.00**	**60.76**	**95.95**	**64.86**	**93.24**	**39.19**	**52.70**	**68.92**	**85.14**
Marginal	69.58	86.95	65.21	100	47.82	100	73.91	78.26	65.21	86.95	60.86	100
Small	75	100	75	100	67.5	92.5	67.5	100	20	20	77.5	72.5
Medium	54.54	10	54.54	100	63.33	100	36.36	100	54.54	100	54.54	100
All Blocks	**74.22**	**86.40**	**68.88**	**89.33**	**61.76**	**90.21**	**71.11**	**78.66**	**68.44**	**81.77**	**58.66**	**66.67**
Marginal	85.19	88.88	75.30	85.18	49.38	93.79	79.01	61.70	76.54	90.12	67.89	80.24
Small	68.75	89.55	69.64	90.17	69.64	85.71	71.43	84.82	64.28	70.53	51.78	47.32
Medium	65.61	69.06	49.97	96.87	65.50	96.87	49.98	100.00	62.48	100.00	59.37	100.00

Table 4.29. Time spent by family members in grazing and milking of livestock

(Multiple response/%)

Block/farm size	Grazing of livestock				Milking		
	Male	Female	Children	Open	Male	Female	Children
Pooh Block	**72.03**	**11.78**	**7.35**	**98.53**	**23.69**	**91.04**	**1.47**
Marginal	74	29.67	18.51	100	48.14	81.14	3.70
Small	68.57	-	-	97.14	3.17	97.14	-
Medium	83.33	-	-	100	33.33	100	-
Kalpa Block	**97.59**	**0.00**	**0.00**	**100.00**	**34.93**	**91.57**	**9.64**
Marginal	100	-	-	100	48.38	100	-
Small	94.59	-	-	100	24.32	81.08	21.62
Medium	100	-	-	100	33.33	100	-
Nichar Block	**94.59**	**21.62**	**10.81**	**100.00**	**40.54**	**100.00**	**0.00**
Marginal	100	26.08	-	100	39.13	100	-
Small	90	25	20	100	45	100	-
Medium	100	-	-	100	27.27	100	-
All Blocks	**88.88**	**10.67**	**5.78**	**99.56**	**33.38**	**94.18**	**4.00**
Marginal	91.33	17.30	6.17	100.00	45.67	93.71	1.23
Small	84.82	8.93	7.14	99.11	25.10	92.86	7.14
Medium	96.87	0.00	0.00	100.00	31.25	100.00	0.00

CHAPTER FIVE

Characteristics of Farming Systems

Determinants of Farming System

The farmers in order to make out a living from the farms allocate the quantities of various basic inputs like land, labour, capital and management. The allocation is made to main production processes like crop, livestock and off-farm enterprises etc. according to his existing knowledge which maximises his profits. Such allocations are also determined by the agro-climatic and better ecological considerations of the area which together form the particular farming system. The emergence and operation of the farming system is the result of various endogenous and exogenous factors. The endogenous factors are within the control of farmers which can be manipulated keeping in view the managerial ability to achieve the desired farm goals, land, labour, capital, management and the production mechanism of farm components. Exogenous factors such as institution, socio-economic and cultural environment, infrastructure facilities and availability of common property resources are beyond the conrtrol of the individual farmer.

Identification of Farming System

In the study of farming system inter-relationship between farm enterprises and household activities are studied. In the present study the pattern of income generation from various sub-systems has been taken as an indicator of a particular farming system.

The relative contribution of different sub-systems towards total farm family income for different blocks and for district Kinnaur has been presented in Tables 5.1 to 5.4.*

A glance at Table 5.1 reveals that in Pooh block, fruit crops account for 40.49 percent of the total family income which is followed by livestock which generates 27.78 percent of the income. The share of agricultural enterprises in total income is 7.14 percent, rest of the income is coming from off-farm and forestry sources. The situation in Kalpa block (Table 5.2) is different where the largest source of income (57.54%) remain fruit crops; the second largest is observed to be CPRs (20.19%). The agricultural crops are generating only 9.54 percent of the total family income. About 13.57 percent of the total income is seen to be coming from off-farm activities. The situation prevailing in Nichar block has been presented in Table 5.3 wherein it may be seen that the fruit crops retain its first position by generating about 50.45 percent of the total income, agriculture crops and CPRs are both generating about 26.35 percent of the total income. The overall picture which emerges for district Kinnaur as a whole has been presented in Table 5.4. It may be seen that about 85.58 percent of the income is generated by agricultural enterprises and the rest by off-farm activities. The study of different sub-systems reveals that the largest share of total family income is received from fruit crops (49.44%). This is followed by contribution of CPRs (14.34%), agricultural crops (9.73%) and livesotck (8.78%). This clearly indicates that the prevailing farming system is a fruit crop-based and much attention is required towards this particular sub-system if the income of the farming families is to be increased. Similarly, the management and development of CPRs and animal husbandry is of prime importance.

Crop income is observed highest in Kalpa block followed by Pooh and Nichar with Rs. 1786, 1683 and 1614 per bigha

---

*See Tables 5.1 to 5.11 at the end of this chapter.

respectively. Per capita income is highest in Pooh block followed by Kalpa and Nichar. Average farm size in Pooh block is 19.89, in Kalpa 21.86 and in Nichar block 19.46 bigha per household. Number of livestock at overall level is lowest in Kalpa block. Per capita income of the study area comprises favourably with state and national figure. According to quick estimates, per capita income of Himachal Pradesh in 1995-96 stood at Rs. 8747 which is less than the Kinnaur district. Per capita income in Kinnaur as a whole is observed at Rs. 8828 whereas national level per capita income of India is Rs. 10,771 (1996-97).

Off-Farm Income

The previous tables reveal that the share of off-farm activities is around 14.42 percent in district Kinnaur. Its further distribution according to actual activities has been presented in Table 5.5. The table reveals that in Pooh block the largest source of off-farm activities is government service from which about Rs. 5160/family/year are received and this is distantly followed by horticulture labour and business. In Kalpa block the service generates about Rs. 4889/family/year and is followed by pension (Rs. 705) and business (Rs. 607). In Nichar block the top position is retained by service (Rs. 6225) and is followed by pension and horticulture. In district Kinnaur the service is top earner activity among the off-farm activities accounting for about Rs. 5410/family/year and is followed by pension (Rs. 704), horticulture labour (Rs. 697) and business (Rs. 530). This indicates that government service is most important off-farm income generating activity in district Kinnaur. The other not so important activities are rural artisan, non-farm labour and agricultural labour.

Employment Generation

The employment generation from different activities on the farm is an important aspect for tackling the problem of under and disguised unemployment in the farming sector. The analysis in this regard reveals that livestock is generating 219 mandays followed by horticulture 153 mandays and CPRs generating about 120 mandays/year in Pooh block. In total, the farm activities are able to generate about 612 mandays/year (Table 5.6). The employment generation in Kalpa block has been presented in Table 5.6 wherein it may be seen that about 707 mandays of employment are generated on the farms. In this block also the pattern of Pooh block is repeated and livestock-rearing is most important in this respect and is generating 294 mandays/year. Horticulture is next important sub-system in this respect and is generating about 191 mandays followed by agriculture and CPRs which account for 111 and 91 mandays. In Nichar block the employment generation is highest among all the three blocks and is about 755 mandays. The employment generation from rearing of livestock increases in this block to about 343 mandays. The horticulture and CPRs are responsible for generation of about 161 and 153 mandays each. The composite picture of district Kinnaur has been presented in Table 5.6 and it reveals that agricultural activities are generating about 96 mandays on an average farm. The livestock sub-system is most important in this respect and generating about 288 mandays/year. Horticulture is observed to be generating about 170 mandays followed by CPRs 120 mandays. The agriculture and farm forestry are other employment generating activities on the farm.

Credit Structure

The farmers can not do without a provision of credit which they require for various activities and inputs like modern inputs, land development, marketing or simply for consumption. The situation in district Kinnaur is not different. The farmers are seen to be taking loans for different purposes like consumption, land improvement, marketing of products and plantation etc. The various sources for taking such loans are State Co-operative Bank, Land Development Bank, Commercial Banks, Agricultural and Horticultural Departments etc. The block-wise details of amount borrowed have been presented in Table 5.7. The table reveals that consumer loans like house and vehicle loans are the main purpose for which largest amount of Rs. 8523 per farm/year is borrowed in district Kinnaur. This is also the main purpose in individual block as well. Thus, the credit is mainly availed for unproductive purposes. The next important purpose is observed to be plantation of fruit plants. For this purpose Rs. 5297 per farm/year are borrowed in district Kinnaur. This purpose also has second position in all the blocks. About Rs. 2432 per farm/year are taken as loan for marketing of fruits and about Rs. 1320 per farm/year for improvement of own land. Giving priorities to unproductive purposes is not a good sign and this practice needs checking. The importance should be given to marketing or land improvement in this regard.

Viability of Farms

The basic purpose of farm production and undertaking of the off-farm activities is that such an income should be able to sustain the farming families. It is desirable that after meeting out the consumption requirements, the family should also be able to save something for investment on the farms for further improvement. The results of the analyses in this regard have been presented in Tables 5.8 to 5.10 for individual blocks and in Table 5.11 for district Kinnaur as a whole. Table 5.8 reveals that per farm annual income in Pooh block is about Rs. 70,291 per farm/year out of which Rs. 61,724 per farm/year is derived from agricultural activities and the rest is off-farm income. At overall level an average family was spending about Rs. 20,123 on food items and about Rs. 32,922 on non-food items every year. This is a positive indication implying that the farm levels in Pooh block are viable, however, if individual size categories are considered the medium farms in this block are found less viable as compared to marginal and small farms.

The annual per farm income from all sources in Kalpa block is observed to be about Rs. 58,189 of which about Rs. 20,472 are spent on food items and Rs. 29,775 on non- food items. Thus, the total expenditure of Rs. 50,247 is less than the total income. This means that the farms in Kalpa block are viable but the medium category of farms are in better condition. In complete contradiction to Pooh block, in Kalpa block marginal and small categories of farms are also viable.

In Nichar block (Table 5.10) a deficit of about Rs. 2144/farm/ year is observed at marginal level, clearly indicating that these farms are not viable. The condition of small and medium farms is particularly bad as compared to Pooh block. The small farms are observed to be viable as indicated by a surplus of about Rs.1868 year/farm.

The picture as emerged in Nichar block is repeated in district Kinnaur. The small farms have a surplus of about Rs. 7417/farm/ year and marginal and medium farms have a surplus of about Rs. 8744 and 11,829/farm/year respectively. The farms at overall level are observed to be having a small surplus of Rs. 8520/farm/ year. It may, therefore, be concluded that there is a need for enhancing the farming income in this district so that in addition to meeting out the consumption expenditure, the farmers are able to save for future investments on the farm; which are required for improving the income generation and hence the standard of living of the farming families. The high consumption expenditure needs curbing in this respect and the credit should be directed towards production purposes.

Table 5.1. Source-wise annual net income (per household) from various farm and off-farm activities of Pooh block

(Net income/Rs./Annum)

Source of income	Marginal	Small	Medium	Overall
1. Income (Rs./Family)				
(A) Farm income				
i) Agricultural crops	2663	5408	13359	5019
	(4.53)	(7.44)	(12.29)	(7.14)
ii) Fruit crops	17652	34792	40142	28458
	(30.04)	(47.88)	(36.94)	(40.49)
iii) Livestock	24284	24538	27238	19528
	(41.33)	(20.01)	(25.06)	(27.78)
iv) Farm forestry	917	2292	1515	1677
	(1.56)	(3015)	(1.39)	(2.38)
v) CPRs	6776	7137	7684	7042
	(11.53)	(9.82)	(7.07)	(10.02)
Sub total (A)	**52292**	**64167**	**89938**	**61724**
	(88.99)	**(88.30)**	**(82.76)**	**(87.81)**
(B) Off- farm income	6468	8502	18733	8567
	(11.01)	(11.70)	(17.24)	(12.19)
Total income (A+B)	**58760**	**72669**	**108671**	**70291**
	(100)	**(100)**	**(100)**	**(100)**
2. Crop income (Rs./bigha)	2306	2002	779	1683
3. Per capita income (Rs.)	12066	11952	8806	11429
4. Average farm size (bigha)	8.81	20.88	68.66	19.89
5. Family size (No.)	4.87	6.08	12.34	6.15
6. No. of livestock	55	42	108	53
7. Sample households (No.)	27	35	6	68

Table 5.2. Source-wise annual net income (per household) from various farm and off-farm activities of Kalpa block

(Net income/Rs./Annum)

Source of income	Marginal	Small	Medium	Overall
1. Income (Rs./Family)				
(A) Farm income				
i) Agricultural crops	3225 (7.15)	5501 (9.94)	10489 (11.37)	5552 (9.54)
ii) Fruit crops	22409 (49.69)	34908 (63.06)	52853 (57.30)	33483 (57.54)
iii) Livestock	-47 (-10)	-5668 (-10.24)	369 (.40)	-2478 (-4.26)
iv) Farm forestry	2005 (4.45)	2015 (3.64)	1874 (2.03)	1985 (3.41)
v) CPRs	10467 (23.21)	10289 (18.59)	17997 (19.51)	11749 (20.19)
Sub total (A)	**38059** **(84.39)**	**47045** **(84.99)**	**83582** **(90.61)**	**50291** **(86.43)**
(B) Off- farm income	7038 (15.61)	8309 (15.01)	8661 (9.39)	7898 (13.57)
Total Income (A+B)	**45097** **(100)**	**55354** **(100)**	**92243** **(100)**	**58189** **(100)**
2. Crop income (Rs./bigha)	2574	2038	1231	1786
3. Per capita income (Rs.)	7996	9134	8686	8646
4. Average farm size (bigha)	9.96	19.83	51.46	21.86
5. Family size (No.)	5.64	6.06	10.62	6.73
6. No. of livestock	20	19	20	19
7. Sample households (No.)	31	37	15	83

Table 5.3. Source-wise annual net income (per household) from various farm and off-farm activities of Nichar block

(Net income/Rs./Annum)

Source of income	Marginal	Small	Medium	Overall
1. Income (Rs./Family)				
(A) Farm income				
i) Agricultural crops	3308	6850	12492	6588
	(9.40)	(13.75)	(15.77)	(13.39)
ii) Fruit crops	16882	24386	42934	24811
	(47.99)	(48.97)	(54.20)	(50.45)
iii) Livestock	-837	2339	-38	553
	((-2.38)	(4.70)	(0.05)	(1.12)
iv) Farm forestry	2416	1919	2162	2110
	(6.87)	(3.85)	(2.73)	(4.29)
v) CPRs	4478	5618	10758	6028
	(12.73)	(11.28)	(13.58)	(12.96)
Sub total (A)	**26247**	**41112**	**68308**	**40090**
	(74.61)	**(82.55)**	**(86.24)**	**(81.51)**
(B) Off- farm income	8931	8689	10901	9093
	(25.39)	(17.45)	(13.76)	(18.49
Total income (A+B)	**35178**	**49801**	**79209**	**49183**
	(100)	**(100)**	**(100)**	**(100)**
2. Crop income (Rs./bigha)	2037	1582	1445	1614
3. Per capita income (Rs.)	6844	6418	9200	6957
4. Average farm size (bigha)	9.91	19.75	38.36	19.46
5. Family size (No.)	5.14	7.76	8.61	7.07
6. No. of livestock	43	38	24	37
7. Sample households (No.)	23	40	11	74

Table 5.4. Source-wise annual net income (per household) from various farm and off-farm activities of Kinnaur district

(Net income/Rs./Annum)

Source of income	Marginal	Small	Medium	Overall
1. Income (Rs./Family)				
(A) Farm income				
i) Agricultural crops	3061 (6.54)	5954 (10.13)	11716 (13.05)	5732 (9.73)
ii) Fruit crops	19254 (41.11)	31114 (52.92)	47060 (52.43)	29112 (49.44)
iii) Livestock	7839 (16.74)	3515 (5.98)	4235 (4.72)	5170 (8.78)
iv) Farm forestry	1759 (3.76)	2067 (3.52)	1906 (2.12)	1933 (3.28)
v) CPRs	7536 (16.09)	7635 (12.99)	13575 (15.13)	8445 (14.34)
Sub total (A)	**39449** **(84.23)**	**50285** **(85.53)**	**78492** **(87.45)**	**50392** **(85.58)**
(B) Off- farm income	7385 (15.77)	8504 (14.47)	11260 (12.55)	8493 (14.42)
Total income (A+B)	**46834** **(100)**	**58789** **(100)**	**89752** **(100)**	**58885** **(100)**
2. Crop income (Rs./bigha)	2334	1865	1171	1701
3. Per capita income (Rs.)	8938	8814	8756	8828
4. Average farm size (bigha)	9.56	19.88	50.18	20.48
5. Family size (No.)	5.24	6.67	10.25	6.67
6. No. of livestock	38	33	38	35
7. Sample households (No.)	81	112	32	225

Table 5.5. Annual off-farm income of sample households of Kinnaur district of H.P.

(Rs./Farm/Annum)

Block/farm size	Source and income								Total
	Govt. job	Agri. labour	Horti. labour	Livestock labour	Trade/ business	N.F. labour	Rural artisan	Pension	income
Pooh Block	**5160.47**	**285.84**	**867.39**	**350.75**	**635.73**	**270.66**	**300.43**	**696.06**	**8567.32**
Marginal	3277.3	349.80	816.90	230.7	719.80	315.90	378.88	378.7	6467.90
Small	5173.45	285.50	917.90	445.9	556.96	218.70	291.42	608.57	8502.4
Medium	13559.0	–	799.90	335.9	716.90	370.18	–	2634.5	18732.57
Kalpa Block	**4888.67**	**108.24**	**551.70**	**511.3**	**606.72**	**106.78**	**419.15**	**705.53**	**7898.09**
Marginal	4232.25	289.80	567.90	431.80	620.90	285.90	483.54	126.03	7038.12
Small	4870.91	–	761.80	551.8	560.80	–	535.13	1028.94	8309.38
Medium	6289.06	–	–	575.7	690.70	–	–	1105.4	8660.86
Nichar Block	**6225.05**	**176.48**	**704.76**	**150.67**	**345.63**	**198.27**	**581.49**	**710.69**	**9093.04**
Marginal	5221.69	567.8	890.80	239.90	437.6	637.90	622.47	312.60	8930.76
Small	6184.0	–	791.60	140.8	387.8	–	467.57	717.5	8689.27
Medium	8472.27	–	–	–	–	–	910.05	1518.27	10900.59
All Blocks	**5410.33**	**184.36**	**697.45**	**344.17**	**529.62**	**186.40**	**436.66**	**704.36**	**8493.35**
Marginal	4194.89	388.74	742.59	310.28	601.82	395.85	488.10	263.23	7385.49
Small	5434.41	89.22	821.22	371.92	497.81	68.34	434.84	786.35	8504.12
Medium	8402.65	–	149.98	332.84	458.18	69.41	312.83	1534.03	11259.93

Table 5.6. Annual mandays (per farm) worked out in different activities

(Mandays)

Activities	Marginal	Small	Medium	All
Pooh Block				
Livestock	233	193	309	219
Horticulture	73	174	385	153
Agriculture	23	126	288	98
Farm forestry	22	20	26	22
CPRs	131	107	141	120
Total	482	620	1149	612
Kalpa Block				
Livestock	262	313	315	294
Horticulture	91	201	375	191
Agriculture	22	113	298	111
Farm forestry	19	20	21	20
CPRs	105	76	99	91
Total	499	723	1108	707
Nichar Block				
Livestock	364	325	359	343
Horticulture	99	162	289	161
Agriculture	15	68	241	77
Farm forestry	19	22	19	21
CPRs	183	150	102	153
Total	680	727	1010	755
Kinnaur District				
Livestock	281	280	328	288
Horticulture	85	179	347	170
Agriculture	20	101	277	96
Farm forestry	20	21	21	21
CPRs	136	112	108	120
Total	542	693	1001	695

Table 5.7. Loans and credit structure of sample farm households

(Rs./Farm/Annum)

Block/farm size	Purpose					Total
	Vehicle purpose	Renovation of own land	Commercial purpose (Mkt. etc.)	Plantation of fruit, plants	For contouring the field	loan
Pooh Block	**8205.87**	**1509.07**	**2303.55**	**6057.56**	**995.78**	**19071.83**
Marginal	4518.5	1365	2167.5	5682	320.7	14053.7
Small	7742.85	1428.57	2375.90	6914.28	1418.70	19880.3
Medium	27500	2627	2493.70	2750	1566.66	36937.36
Kalpa Block	**8639.76**	**1009.80**	**2242.31**	**6162.65**	**776.79**	**18831.30**
Marginal	519.35	322.58	2137.8	5000	217.8	8197.53
Small	12972.97	1125.8	2270.9	5108.1	1087.59	22565.36
Medium	14733.33	2143.9	2387.8	11166.66	1165.40	31597.09
Nichar Block	**8682.43**	**1495.03**	**2762.97**	**3628.37**	**901.53**	**17470.34**
Marginal	5869.56	1164.80	2870	2826.08	189.9	12920.34
Small	8937.5	1506.8	2650	3962.5	1182.5	18239.3
Medium	13636.36	2142.72	2950	4090.90	1367.8	24187.78
All Blocks	**8522.66**	**1320.28**	**2432.06**	**5297.39**	**884.00**	**18456.39**
Marginal	3371.60	909.20	2355.61	4610.05	244.18	11490.63
Small	9897.32	1365.49	2439.11	5263.39	1224.96	20181.86
Medium	16750	2234.08	2600.91	7156.24	1310.21	30051.44

Major source of credit: (i) State Co-operative Bank, (ii) Land Development Bank, (iii) Commercial Banks, (iv) Agricultural Depart-ment and, (v) Horticultural Department.

Table 5.8. Actual consumption expenditure and viability of farms of Pooh block

(Rs./Per farm/Year)

	Marginal	Small	Medium	All
(A) Food items				
Wheat, rice, sugar & edible oil	11735	13821	17307	13450
Pulses (mash, gram, lentil, mix, mong, rongi etc.)	812	925	1454	927
Spices, tea/coffee, tobacco, soap, detergent, meat & wine etc.	3849	4720	8650	4721
Milk, ghee, eggs, vegetables fruits & medicines etc.	823	885	1920	1025
Sub total (A)	**17219**	**20351**	**29331**	**20123**
(B) Non-food items				
Clothing, footwears, ritual, ceremonies, LPG, education, tel. & electricity, panchayat rent, land revenue, repair home & appliances etc.	20301	31804	56670	29431
Horticulture medicines, tree spray, oil, fertilisers, equipments & concentrate & veterinary exp. on livestock	2756	3546	6483	3491
Sub total (B)	**23057**	**35350**	**63153**	**32922**
Grand total (A+B)	**40276**	**55701**	**92484**	**53045**
Total income from livestock, horticulture, agriculture, CPRs & farm forestry	52292	64167	89938	61724
Off-farm income	6468	8502	18733	8567
Total farm income	**58760**	**72669**	**108671**	**70291**
Surplus/deficit	18484	16968	16187	17246

Table 5.9. Actual consumption expenditure and viability of farms of Kalpa block

(Rs./Per farm/Year)

	Marginal	Small	Medium	All
(A) Food items				
Wheat, rice, sugar & edible oil	11470	13824	16513	13223
Pulses (mash, gram, lentil, mix, mong, rongi etc.)	706	736	1588	879
Spices, tea/coffee, tobacco, soap, detergent, meat & wine etc.	4185	5817	6800	5385
Milk, ghee, eggs, vegetables fruits & medicines etc.	396	845	2545	985
Sub-total (A)	**16757**	**21222**	**27446**	**20472**
(B) Non-food items				
Clothing, footwears, ritual, ceremonies, LPG, education, tel. & electricity, panchayat rent, land revenue, repair home & appliances etc.	17276	26905	43957	26390
Horticulture medicines, tree spray, oil, fertilisers, equipments & concentrate & veterinary exp. on livestock	2634	3363	4993	3385
Sub-total (B)	**19910**	**30268**	**48950**	**29775**
Grand total (A+B)	**36667**	**51490**	**76396**	**50247**
Total income from livestock, horticulture, agriculture, CPRs & farm forestry	38059	47045	83582	50291
Off-farm income	7038	8309	8661	7898
Total farm income	**45097**	**55354**	**92243**	**58189**
Surplus/deficit	8430	3864	15847	7942

Table 5.10. Actual consumption expenditure and viability of farms of Nichar block

(Rs./Per farm/year)

	Marginal	Small	Medium	All
(A) Food items				
Wheat, rice, sugar & edible oil	13994	13114	15231	13702
Pulses (mash, gram, lentil, mix, mong, rongi etc.)	684	528	1958	730
Spices, tea/coffee, tobacco, soap, detergent, meat & wine etc.	3817	5443	6391	5078
Milk, ghee, eggs, vegetables fruits & medicines etc.	399	857	2253	922
Sub-total (A)	**18894**	**19942**	**25834**	**20432**
(B) Non-food items				
Clothing, footwears, ritual, ceremonies, LPG, education, tel. & electricity, panchayat rent, land revenue, repair home & appliances etc.	16305	24949	40874	24629
Horticulture medicines, tree spray, oil, fertilisers, equipments & concentrate & veterinary exp. on livestock	2123	3042	4531	2977
Sub-total (B)	**18428**	**27991**	**45405**	**27606**
Grand total (A+B)	**37322**	**47933**	**71239**	**48038**
Total income from livestock, horticulture, agriculture, CPRs & farm forestry	26247	41112	68308	40090
Off-farm income	8931	8689	10901	9093
Total farm income	**35178**	**49801**	**79209**	**49183**
Surplus/deficit	-2144	1868	7970	1145

Table 5.11. Actual consumption expenditure and viability of farms of Kinnaur district

(Rs./Per farm/year)

	Marginal	Small	Medium	All
(A) Food items				
Wheat, rice, sugar & edible oil	12275	13416	16539	13449
Pulses (mash, gram, lentil, mix, mong, rongi etc.)	707	712	1656	844
Spices, tea/coffee, tobacco, soap, detergent, meat & wine etc.	3969	5340	7006	5083
Milk, ghee, eggs, vegetables fruits & medicines etc.	600	862	2328	976
Sub-total (A)	**17551**	**20330**	**27529**	**20352**
(B) Non-food items				
Clothing, footwears, ritual, ceremonies, LPG, education, tel. & electricity, panchayat rent, land revenue, repair home & appliances etc.	18009	27737	45281	26730
Horticulture medicines, tree spray, oil, fertilisers, equipments & concentrate & veterinary exp. on livestock	2530	3305	5113	3283
Sub-total (B)	**20539**	**31042**	**50394**	**30013**
Grand total (A+B)	**38090**	**51372**	**77923**	**50365**
Total income from livestock, horticulture, agriculture, CPRs & farm forestry	39449	50285	78492	50392
Off-farm income	7385	8504	11260	8493
Total farm income	**46834**	**58789**	**89752**	**58885**
Surplus/deficit	8744	7417	11829	8520

CHAPTER SIX

Economics of Existing Farming Systems

This section deals with the economics of existing farming systems of Kinnaur. Gross income, labour use pattern, expenditure on variable inputs and returns over variable cost have been discussed for the farm as a whole unit as well as for different enterprises.

Net Returns

The annual net returns from various farm components under different farm categories are summarised in different tables.

Agricultural Crops

The results of analysis have been presented in Tables 6.1 to 6.5.* It may be seen from the table that by cultivating maize, farmers are incurring a loss of about Rs. 118/farm in the district. The positive net return of about Rs. 103/farm is observed in Nichar block. In other two blocks the farmers are in deficit in maize production. Per farm net income from wheat cultivation is Rs. 244 in Nichar and Rs. 288/farm in Pooh block. The cultivation of barley resulted in a net loss of about Rs. 116/farm at overall level. The cultivation of miscellaneous crops like koda, chollai, phapra and ogla etc. is again a losing preposition in all the blocks. Rajmash, potato and pea are noted to be significant contributors to farm income. At overall level the cultivation of rajmash resulted

---

*See tables 6.1 to 6.29 at the end of this chapter.

in a net inflow of about Rs. 3869/farm, while potato cultivation stood at Rs. 1359/farm. However, during the survey year, peas crop was not normal because of heavy rains. In aggregate, agriculture sector is running in surplus in the district (Rs. 5732/ farm).

In agriculture sub-sector maximum net return per farm is from cash crops like rajmash and potato. People are also showing more interest in the production of cash crops. But due to some social obligations and traditions, they are still producing some particular type of crops like barley, ogla and phapra. Barley is generally used in religious ceremonies and also used as feed to the livestock.

Farm Yard Manure (FYM)

It is observed that FYM consumption in study area is very high because the area is situated in dry zone. But the consumption of chemical fertilizers in agriculture is very low. The detail of consumption and allocation of FYM is presented in Table 6.6.

Sources of Input

The inputs considered in the analyses are seeds and fertilizers. The various sources of these inputs are observed to be department of agriculture, horticulture, block office and market (Table 6.7).

Seeds: The main source of seeds is the department of agriculture, more than 90 percent of the farmers are procuring seeds from this source. The local source of seeds is the next in importance with 51.56 percent farmers at overall level are using this. About 41 percent and 36.44 percent farmers at overall level were obtaining seeds from block office and open market respectively (Table 6.7).

Fertilizers: The main source of fertilizers procurement is societies, and according to 65 percent farmers on an average they are using this source. About 49.30 percent, 31.56 percent and 4.44 percent respondents are also observed to be buying fertilizers from open market, horticulture department and agriculture department respectively in district Kinnaur.

Changes and Opinions about Farming System

The farmers' perception about the modernisation of farming and its various economic advantages have been obtained as the multiple responses and are presented in Table 6.8. It may be seen that about 33 percent of farmers feel that there has been a modernisation of agriculture during the last 5 years. About 77 percent farmers reveal that there has been modernisation of horticultural sector. About 71 percent farmers thought that the horticultural sector still requires more attention. This is because about 73 percent respondents thought that horticulture cultivation is more profitable in district Kinnaur under the given agro-climatic and agro-ecological conditions. The profitability of horticultural sector in this region is the main cause which motivated the government according to 84 percent respondents at overall level, for paying more attention to this particular sector.

The responses of farmers regarding the profitability of different farming systems were also recorded and the results of analysis indicate that 80 percent farmers at overall level thought that agriculture enterprise is a loss-making proposition. About 83 percent farmers opined that horticulture is the venture which results in net profits for the farmers. Similarly about 63 percent farmers thought that livestock is a profit-making activity in the area. It may, therefore, be concluded that horticulture and livestock, the two major components of farming system, are profitable under the local conditions. The government has also realised this and paid more attention to horticulture sector. The animal husbandry, according to respondents has not been paid as much attention as it deserves.

FRUIT CROPS

Initial Costs of Plantation

The initial cost of apple plantation is presented in Table 6.9 wherein it may be seen that, at overall level, initial cost is Rs. 35.53 per plant, of which Rs. 12.53 is on digging of pits and Rs. 4.72 on filling it and Rs. 12.73 are spent as cost of plant.

The expenditure on FYM, fertilizers, irrigation, fencing and staking (plant support) etc. is Rs. 5.56. The highest initial cost of plantation is found to be in Pooh block (Rs. 38.14) followed by Kalpa (Rs. 35.36) and Nichar (Rs. 33.32). Invariably the highest cost component is cost of plant followed by digging of pits.

Marketing Cost

The cost of marketing of apples includes all costs involved in taking the fruit from the farm to the market, which in present case is Delhi. This includes all costs on packing case (box), picking, grading, packing, transportation, taxes etc. The results of this analysis have been presented in Table 6.10. It may be seen that average cost of marketing of apples is Rs. 120 per box in Delhi. The commission is the largest cost component (Rs. 32.75/box) followed by cost of transportation (Rs. 32.50/box). Rs. 30.63 are paid for empty box in the study area. Another major cost component is packing of fruits for which Rs. 12.13/box had to be paid. The cost of marketing was observed to be highest in Pooh block (Rs. 132.14/box) mainly due to higher cost of transportation, followed by Kalpa and Nichar, where it is almost equal. There is minor variation in cost but the trend remained almost same in all the blocks.

Sources of Plant Protection Material (PPM) and Machinery/Equipment

The department of agriculture, horticulture, block office and open market are the sources of plant protection material. The details in this regard, have been presented in Table 6.11 wherein it may be seen that main source of plant protection material is department of horticulture; according to 95.61 percent farmers they obtained plant protection material (PPM) from this source. The percentage of respondents who obtained plant protection material from this source is 98.57 percent in Pooh, 92.70 percent in Kalpa and 96.07 percent in Nichar block respectively. Another major source is block office according to 49.68 percent of respondents in the study area. The table also depicts sources of horticulture machinery/equipment. The analysis reveals that about 85.33 percent of respondents at overall level, depend on the department of horticulture, 25.80 percent on open market and 17.66 percent on local block office for meeting these requirements.

Farm Yard Manure (FYM)

The availability of FYM and its utilization for different crops is presented in Table 6.12. It may be seen from the table that on an average 68.40 qtls. of FYM is available on each farm in the district. This quantity is 80.03 qtls., 62.74 qtls. and 64.05 qtls. in Pooh, Kalpa and Nichar blocks respectively. About half of this quantity is used for apples and the rest for other horticultural and agricultural crops. The proportion of FYM used in horticulture and agriculture is about 51.77 percent and 48.23 percent in the district.

Fertilizers

It is seen that 1.65 qtls. of chemical fertilizers are being consumed on each farm at overall level (Table 6.13). Majority of this quantity is Cane & Urea (0.78 qtls.) followed by Superphosphates (0.47 qtls.) and Potash (0.40 qtls.) in the district. This pattern is same in all the blocks. The highest consumption of fertilizers is in Kalpa (2.11 qtls.), followed by Nichar (1.51 qtls.) and Pooh block (1.23 qtls.) respectively. The fertilizers consumption is seen to increase with farm size.

Marketing Methods

Details about marketing methods in the study area is presented in Table 6.14. Being a geographically tough area, marketing of fruits is a very complicated process. Farmers have to face many problems like transportation, price risk etc. So, they sell their produce generally to village merchants, commission agents and local forwarding agents.

Net Income from Horticulture

The net income generated by cultivation of horticultural crops has been presented in Table 6.15 wherein it may be seen that a gross income of Rs. 44,933 is generated at overall level of which Rs. 15,821 is total expenditure and thus a net income of Rs. 29,112 per year per farm is received. Among the horticulture crops the cultivation of apple is the main source of income of Rs. 40152.38 per farm/year which is followed by almonds which generates only Rs. 2737 per farm/year. This pattern is seen to exist in all the blocks. The net income from fruit crops in Pooh block is Rs. 28,459, in Kalpa Rs. 33,483 and in Nichar Rs. 24,811 per farm/year. The analysis clearly projects apple as the dominant source of income for the people in the study area. Almond production is maximum in Pooh block due to dry climatic conditions. Grapes which are used generally to make local wine (Angoori) is mainly produced in Kalpa block.

LIVESTOCK

Livestock Management during Migration

During migration the flock owners generally send a family member or servant with their flocks. Those small flock owners who do not possess grazing permits in forest lands may accompany the migrants, in group by posing as servants. The servants of the migrants are paid either in cash or kind with a few sheep and goat every year. In this way, after 2-3 years' service they become the owner of a large flock. Some flock owners send their flocks with other shepherds after making arrangements for payment of relevant expenses. The expenses incurred during the migration include the fee and taxes paid to forest department or village panchayats for grazing permits, cost of feed, food expenditure of the accompanying persons etc. However, the animals that are lost or die are debited to the account of the owners. Animal identification is done by notching the ears in different patterns with locally made tools.

During migration, both upward and downward, the shepherds

may visit their home villages. The animals remain in open all the year round and are never penned. Generally it is noticed that the member of the family who has accompanied the flock during winter does not have to go during summer migration unless there is a great need. Due to the short distance of the pastures the persons keep on changing. In alpine region, there are vast undulating grasslands area, so, the animals are let free to move and graze on their own for months together without any special care.

Trained shepherd dogs are used for guiding the flocks. Watch dogs are also used by some flock owners. On an average 3-4 dogs for the protection of sheep and goats from the wild animals and thieves accompany the flock. These dogs play an important role in flock management and are quite understanding and intelligent.

In some areas of Kinnaur, sheep and goats are used for transporting rations and other agricultural produce. Male animals used for transportation are locally called 'laadoo' and can carry about 8-10 kg. of load.

The average number of different type of livestock migrated during winters (out of district) have been presented in Table 6.16 alongwith other details. The table reveals that average migration period is 4.24 months and sheep and goats are the main livestock migrated. The percentage of goat and sheep migration is 41.77 percent and 45.77 percent respectively. With this 41.77 percent donkeys and 31.11 percent mules are taken along to carry the provisions. It is seen that 68.81 percent flocks are accompanied by the owners and about 11 percent flocks are looked after by hired hand during migration in the district. Further nearly 20 percent flocks are looked after by the contractor. This pattern is also seen in the individual blocks as well.

During summer and rainy season, livestock is also migrated to local high hills (*kandas*) for about 7 months. The big advantage of grazing in these *kandas* is their less expensiveness because there is no need of supervision during this migration. The detail in this regard is given in Table 6.17.

Income from Sheep and Goats

Keeping in view the importance of sheep and goats, the income generation from these has been dealt separately and presented in Tables 6.18 to 6.20. It may be seen from Table 6.18 that total income of Rs. 5167.00 farm/year is generated by sheep at overall level. Out of this, sale in market for meat purpose accounts for Rs. 3292.13, domestic use for meat purpose Rs. 732.56 per year and sale of wool Rs. 1142.64 per farm/year in the district. The income from sheep (Rs. 7295.39) is the highest in Pooh followed by Nichar (Rs. 5269.81) and Kalpa (Rs. 3332.49) per farm/year respectively. The income derived from rearing goats has been presented in Table 6.19 depicting that at overall level goats generate an income of Rs. 10869.10 per farm/year. This income is mainly the result of sale in market (Rs. 6971.47) followed by sale of wool (Rs. 2081.47) and domestic use (Rs. 1816.15). Like sheep, the highest income from goats is observed in Pooh (Rs. 14678.64) followed by Nichar (Rs. 12,989.44) and the least (Rs. 5857.60) in Kalpa block. The total income from sheep and goats has been presented in Table 6.20 showing that total income from sheep and goats is Rs. 17,062.42 per farm/year in the district.

Feed/Concentrate

In addition to fodder the animals are also fed concentrates for improving productivity. The quantities of concentrates fed per household have been presented in Table 6.21 wherein it may be seen that total quantity of 2.95 qtls. of concentrate is fed to livestock per farm/year in the district. Out of this 1.04 qtls. are fed during winters, 0.93 qtls. in summer and 1.30 qtls. during rainy season/year. The block-wise trend reveals that average quantity per farm was 3.11 qtls. 2.93 qtls., and 2.82 qtls. in Pooh, Kalpa and Nichar blocks respectively. The sources of procurement of the concentrates alongwith their value is presented in Table 6.22 wherein it may be seen that of the 2.95 qtls. of total concentrates fed at overall level, 0.98 qtls. is purchased from open market and the rest 1.96 qtls. is produced at home and these have value of Rs. 784 and 1811 respectively. It may be seen that large proportion of concentrates is made at home. The total value of concentrates fed is Rs. 1990, Rs. 2794 and Rs. 2928 in Pooh, Kalpa and Nichar blocks respectively.

Income from the Livestock

The gross income generated by different types of livestock has been presented in Table 6.23 wherein it may be seen that each household at overall level is receiving a gross income of Rs. 32,473.35 per year and this figure is Rs. 41,607.28 in Pooh, Rs. 24,932.49 in Kalpa and Rs. 32,537.39 in Nichar block respectively. More than 50 percent of their income is derived by rearing sheep and goats. The various costs incurred for generating this income and consequent net income have been presented in Table 6.24. At overall level, per farm net returns are seen to vary inversely with holding size. The highest net income is in Pooh block (Rs. 19,528.54 per farm) followed by Nichar where it is only Rs. 533.61. The farmers of Kalpa are incurring a loss by rearing livestock and this loss is estimated to be Rs. 2477.58 per farm per year.

NET INCOME FROM FARM FORESTRY

The income generated by farm forestry is evaluated and the results of trends have been presented in Table 6.25 wherein every household on an average is seen to be obtaining a gross income of Rs. 3342.70. This income is constituted of Rs. 501.54 from timber, Rs. 1317.22 from fuelwood, Rs. 1260.31 from grazing & grasses and Rs. 263.63 from miscellaneous uses like getting material for fencing and staking of vegetable crops. About Rs. 1409 are spent by each household for deriving this income resulting in a net income of Rs. 1933.27 per farm/year in district Kinnaur. No definite link of income from farm forestry with size of holding can be established. The net income from this enterprise is seen to be Rs. 1677.70 in Pooh, Rs. 1985.43 in Kalpa and Rs. 2109.61 per farm/year in Nichar block respectively.

Net Income from CPRs

The farmers of Kinnaur heavily depend on common property resources to meet their requirements of grass, fuelwood, timber etc. The quantities of these goods extracted are evaluated to find the income from CPRs. The results of analysis have been presented in Tables 6.26 to 6.28. Table 6.26 presents the details regarding gross income generated from major uses of CPRs. The major uses include grazing, fodder, fuelwood and timber. It may be seen from the table that at overall level a gross income of Rs. 14867.61 is generated by each household and major proportion of this is constituted by grazing of animals on CPRs. The income generated from minor uses of CPRs includes collection of chilgoza, herbs etc. and the details have been presented in Table 6.27. It may be seen that each household is receiving a gross income of Rs. 1260.45 from minor uses of CPRs. The largest proportion of this income is derived from collection of chilgoza. The total income from major and minor uses of CPRs alongwith net income after deduction of expenses has been presented in Table 6.28. It may be seen that each household is deriving a net income of Rs. 8444.59 from CPRs. The highest income per household/year is seen in Kalpa block (Rs. 11,748.63), followed by Pooh (Rs. 7041.78) and Nichar block (Rs. 6027.79) respectively.

Mostly the values of CPRs products are evaluated at the local market prices. The forest corporation has fixed prices for soft wood (Rs. 150/qtls.), hard wood (Rs. 181/qtls.) and slippers (Rs. 900-950) respectively.

During this survey information regarding the value of land per bigha was observed which is highest in Pooh block. In the district Kinnaur per bigha cost of crop land and orchard land is Rs. 20,356 and Rs. 53,873 respectively. Farmers also have their own sheds for their agriculture product and livestock shelter. Many farmer have their own store for the storage of foodgrain (Table 6.29).

Table 6.1. Net returns over variable cost in maize and wheat of sample households

(Rs./Farm)

Block/ farm size	Cereal crop						Overall maize and wheat		
	Maize			Wheat					
	Income	Expen-diture	Deficit or surplus	Income	Expen-diture	Deficit or surplus	Total income	Total exp.	Deficit or surplus
Pooh Block	**955.97**	**1189.02**	**-233.05**	**2557.04**	**2268.88**	**288.16**	**3513.01**	**3457.90**	**55.11**
Marginal	427.75	630.50	-202.75	1586.88	1254.5	332.38	2014.63	1885	129.63
Small	929.74	1495.0	-565.26	2811.41	2554.5	256.91	3741.15	4049.5	-308.35
Medium	3485.98	1917.50	1568.48	5438.98	5167.5	271.48	8924.96	7085	1839.96
Kalpa Block	**1028.29**	**1249.88**	**-221.59**	**2827.88**	**2540.64**	**287.23**	**3856.16**	**3790.52**	**65.64**
Marginal	672.56	780	-107.44	1602.60	1287	315.60	2275.16	2067	208.16
Small	1147.04	1365	-217.96	2817.2	2613	204.20	3964.24	3978	-13.76
Medium	1470.55	1937	-466.45	5386.4	4953	433.4	6856.95	6890	-33.05
Nichar Block	**1325.78**	**1223.14**	**102.64**	**2694.00**	**2450.41**	**243.59**	**4019.78**	**3673.55**	**346.23**
Marginal	711.36	638.13	73.23	1486.99	1365	121.99	2198.35	2003.13	195.22
Small	1448.62	1365	83.62	8736.99	2502.5	234.50	4185.61	3867.5	318.12
Medium	2163.81	1930.5	233.31	5061.39	4530.5	530.89	7225.2	6461	764.2
All Blocks	**1104.28**	**1222.69**	**-118.48**	**2701.99**	**2428.83**	**273.16**	**3806.21**	**3651.53**	**154.68**
Marginal	601.97	689.88	-88.08	1564.53	1298.31	266.22	2166.34	1988.20	178.14
Small	1186.84	1405.63	-218.78	2786.74	2555.25	231.49	3973.59	3960.88	12.71
Medium	2086.75	1931.11	155.64	5284.54	4847.98	436.55	7371.29	6779.09	592.19

Table 6.2. Net returns over variable cost in barley and misc. crops of sample households

(Rs./Farm)

Block/ farm size	Cereal crop						Overall barely and misc. crops		
	Barley			Misc. crops					
	Income	Expen- diture	Deficit or surplus	Income	Expen- diture	Deficit or surplus	Total income	Total exp.	Deficit or surplus
Pooh Block	**1854.96**	**2086.02**	**-231.07**	**696.29**	**706.38**	**-10.09**	**2551.25**	**2792.40**	**241.16**
Marginal	477.4	780	-302.6	379.5	385	-5.5	856.90	1165	308.10
Small	2542	2567.5	-25.5	696.9	707	-10.8	3238.9	3274.50	35.60
Medium	4046.20	5154.5	-1108.3	2118.3	2149	-30.7	6164.50	7303.5	1139
Kalpa Block	**2020.23**	**2081.08**	**-60.95**	**903.57**	**916.66**	**-13.10**	**2923.80**	**2997.84**	**118.49**
Marginal	558	494	64	310.5	315	-4.5	868.50	809	59.50
Small	2418	2418	0	703.8	714	-10.2	3121.80	3132	10.2
Medium	4061	4530.50	-469.5	2622	2660	-38	6683	7190.5	507.50
Nichar Block	**2181.23**	**2252.53**	**-71.37**	**903.45**	**918.51**	**-15.07**	**3084.67**	**3170.80**	**113.31**
Marginal	632.40	611.00	21.4	414	420	-6	1046.40	1031	15.40
Small	2498.60	2470	28.60	617.4	630	-12.6	3116	3100	16
Medium	4265.60	4895.5	-628.9	2967	3010	-43	7232.6	7904.50	671.90
All Blocks	**2023.23**	**2138.96**	**-115.79**	**840.88**	**853.72**	**-12.84**	**2864.11**	**2992.64**	**153.86**
Marginal	552.26	622.56	-70.30	362.89	368.15	-5.26	915.15	990.42	129.84
Small	2485.54	2483.29	2.25	670.79	681.81	-11.03	3156.32	3165.10	20.21
Medium	4128.56	4772.22	-644.07	2646.15	2684.50	38.35	6774.71	7457.13	682.42

Table 6.3. Net returns over variable cost in rajmash, potato and peas of sample households

(Rs./Farms)

Block/ farm size	Major cash crops											
	Rajmash			Potato			Peas			Overall		
	Income	Expen-diture	Deficit or surplus	Income	Expen-diture	Deficit or surplus	Income	Expen-diture	Deficit or surplus	Income	Expen-diture	Deficit or surplus
Pooh Blk.	**4261.55**	**1076.35**	**3185.20**	**4115.86**	**2706.62**	**1409.25**	**2452.41**	**2094.71**	**357.71**	**10829.83**	**5877.68**	**4952.15**
Marginal	2340	560	1780	2281	1500	781	1320	1200	120	5941	3260	2681
Small	4773.6	1190	3583.6	4790.10	3150	1640.10	2508	2280	228	12071.7	6620	5451.70
Medium	9921.6	2737	7184.6	8439.7	5550	2889.7	7224	5040	2184	25585.3	13327	12258.3
Kalpa Blk.	**4923.01**	**1360.61**	**3562.40**	**4306.42**	**2831.93**	**1474.49**	**2462.04**	**2238.22**	**223.82**	**11691.47**	**6430.76**	**5260.71**
Marginal	2652	686	1966	2052.90	1350	702.90	1188	1080	108	5892.9	3116	2776.9
Small	4774	1365	3409	4333.9	2850	1483.9	2547.6	2316	231.6	11655.5	6531	5124.5
Medium	9984	2744	7240	8895.90	58.50	3045.9	4884	4440	444	23763.9	13034	10729.9
Nichar Blk.	**6182.66**	**1340.22**	**4842.44**	**3456.95**	**2273.31**	**1183.64**	**0.00**	**0.00**	**0.00**	**9639.61**	**3613.53**	**6026.08**
Marginal	2901.6	630	2271.6	1710.75	1125	585.75	-	-	-	4612.35	1755	2857.35
Small	6552	1351	5201.0	2965.3	1950	1015.3	-	-	-	9517.3	3301	6216.3
Medium	11700	2786	8914	8895.9	5850	3045.9	-	-	-	20595.9	8636	11959.9
All Blks.	**5137.39**	**1268.00**	**3869.39**	**3969.45**	**2610.33**	**1359.11**	**1649.39**	**1458.72**	**190.67**	**10756.23**	**5337.05**	**5419.18**
Marginal	2618.87	628.10	1990.78	2031.78	1336.11	695.67	894.67	813.33	81.33	5545.32	2777.54	2767.78
Small	5408.88	1305.31	4103.57	3987.68	2622.32	1365.36	1625.37	1477.61	147.76	11021.92	5405.24	5616.68
Medium	10562.18	2757.13	7805.05	8810.36	8793.75	3016.61	3643.88	3026.25	617.63	23016.41	11577.13	11439.29

Table 6.4. Net returns after all cost from vegetables (other than peas and potato) of sample households

(Rs./Farm/Annum)

Block/farm size	Income	Expenditure	Deficit/Surplus
Pooh Block	**379.85**	**126.62**	**253.24**
Marginal	240	80	160
Small	450	150	300
Medium	600	200	400
Kalpa Block	**449.64**	**149.88**	**299.76**
Marginal	270	90	180
Small	600	200	400
Medium	450	150	300
Nichar Block	**453.24**	**151.08**	**302.16**
Marginal	360	120	240
Small	450	150	300
Medium	660	220	440
All Blocks	**429.73**	**143.24**	**286.49**
Marginal	285.56	95.19	190.37
Small	499.55	166.52	333.04
Medium	550.31	183.44	366.88

Table 6.5. Net returns over variable cost in agriculture crops of sample households

(Rs./Farm/Annum)

Block/farm size	Total agriculture income	Total agriculture expenditure	Overall agri. surplus/deficit
Pooh Block	**17273.94**	**12254.60**	**5019.34**
Marginal	9052.53	6390	2662.53
Small	19501.75	14094	5407.75
Medium	41274.76	27915.5	13359.26
Kalpa Block	**18921.13**	**13368.99**	**5552.07**
Marginal	9306.56	6082	3224.56
Small	19341.54	13841	5500.54
Medium	37753.85	27264.50	10489.35
Nichar Block	**17197.30**	**10609.28**	**6588.03**
Marginal	8217.10	4909.13	3307.97
Small	17268.91	10418.5	6850.41
Medium	35713.7	23221.5	12492.20
All Blocks	**17856.34**	**12124.56**	**5731.78**
Marginal	8912.53	5851.63	3060.90
Small	18651.38	12697.74	5953.64
Medium	37712.72	25996.78	11715.94

Table 6.6. Consumption of FYM and its allocation by sample farms

(Qtls./%)

Block/ farm size	FYM total qty. qtls.	Horticulture				Agriculture							
		Apple	Al-mond	Apri-cot	Plum	Wheat	Barley	Maize	Raj-mash	Rongi	Peas	Potato	Others
Pooh Block	**80.03**	**41.68**	**5.01**	**4.96**	**0.00**	**14.92**	**12.63**	**3.07**	**11.96**	**0.00**	**2.69**	**1.22**	**1.87**
Marginal	89.33	36.69	7.45	3.61	0	17.12	12.18	5.6	10.98	0	3.99	2.19	0.19
Small	65.43	43.41	3.05	6.27	0	14.63	13.75	1.65	12.17	0	1.37	0.21	3.49
Medium	123.3	54.07	5.41	3.38	0	6.75	8.1	0	15.09	0	4.5	2.7	0
Kalpa Block	**62.72**	**45.20**	**2.70**	**1.10**	**0.17**	**15.33**	**15.49**	**3.74**	**11.47**	**0.00**	**2.12**	**1.63**	**1.04**
Marginal	56.48	32.26	3.18	0	0	18.46	17.43	6.4	12.74	0	4.86	3.24	1.43
Small	69.18	60.94	1.95	1.56	0.39	11.72	12.3	0.39	10.08	0	0	0	0.67
Medium	59.66	33.14	3.57	2.23	0	17.78	19.33	6.49	12.29	0	1.68	2.33	1.16
Nichal Block	**64.05**	**49.60**	**0.23**	**0.00**	**0.00**	**15.61**	**11.41**	**9.10**	**10.45**	**0.61**	**1.64**	**0.61**	**0.75**
Marginal	59.34	35.11	0	0	0	20.57	17.58	10.62	11.36	0	3.3	0.73	0.73
Small	65.13	55.85	0	0	0	14.01	9.21	9.02	9.4	0.77	0.78	0	0.96
Medium	70	57.14	1.56	0	0	11.04	6.49	6.23	12.38	1.29	1.29	2.58	0
All Blocks	**68.39**	**45.58**	**2.59**	**1.90**	**0.06**	**15.30**	**13.28**	**5.30**	**11.28**	**0.20**	**2.13**	**1.17**	**1.20**
Marginal	68.24	34.55	3.70	1.20	0.00	18.61	15.72	7.33	11.76	0.00	4.13	2.18	0.82
Small	66.56	53.64	1.60	2.47	0.13	13.45	11.65	3.87	10.49	0.28	0.71	0.07	1.65
Medium	75.15	45.31	3.22	1.68	0.00	13.40	12.81	5.18	12.85	0.44	2.07	2.49	0.54

Table 6.7. Sources of seeds and fertilisers of sample households

(Multiple response/%)

Block/farm size	Seeds				Fertilisers			
	Agriculture dept.	Block	Local source	Open market	Agriculture dept.	Societies	Horticulture dept.	Open market
Pooh Block	**91.14**	**57.35**	**55.84**	**19.11**	**8.82**	**44.84**	**17.67**	**35.20**
Marginal	92.5	37.03	77.7	7.4	11.11	81.45	44.5	40.5
Small	91.42	77.14	45.7	20	5.71	10.0	-	28.57
Medium	83.33	33.3	16.66	66.6	16.67	83.33	-	50
Kalpa Block	**87.70**	**42.36**	**26.51**	**38.55**	**2.39**	**71.17**	**43.37**	**21.67**
Marginal	91.11	32.25	32.29	19.35	-	22.8	100	-
Small	78.37	59.9	24.3	35.13	-	100	-	21.62
Medium	93.33	20	20	86.66	13.23	100	33.3	66.6
Nichar Block	**97.30**	**25.94**	**75.74**	**50.00**	**2.70**	**75.67**	**31.08**	**93.24**
Marginal	100	30.43	60.86	82.60	-	26.08	100	95.65
Small	100	20.5	97.5	25	-	97.5	-	95
Medium	81.81	36.36	27.7	72.72	18.18	100	-	81.81
All Blocks	**91.89**	**41.49**	**51.56**	**36.44**	**4.44**	**64.69**	**31.56**	**49.30**
Marginal	96.01	33.33	55.54	33.33	3.70	43.28	81.50	40.66
Small	90.17	51.22	57.13	26.78	1.78	70.98	0.00	50.00
Medium	87.50	28.12	22.02	78.11	15.58	96.87	15.61	68.72

Table 6.8. Major changes in last five years in the farming system of sample farms

(Multiple response/%)

Block/ farm size	Major changes in last 5 years					What is your opinion about F/S					
	Mod. in agri-culture	Mod in horti-culture	More attention on horti.	Horti. more beneficial	More atten. by govt. to horti.	Agriculture		Horticulture		Livestock	
						Loss making	Profit making	Loss making	Profit making	Loss making	Profit making
Pooh Block	**24.99**	**70.58**	**74.97**	**59.07**	**83.81**	**86.77**	**13.23**	**14.70**	**85.30**	**33.83**	**66.17**
Marginal	33.3	66.66	88.88	37.03	74.07	96.3	3.7	7.4	92.6	37.03	62.97
Small	20	71.42	68.52	74.78	91.42	82.85	17.15	17.15	82.85	34.3	65.7
Medium	16.66	83.3	50	66.66	83.3	66.7	33.3	33.3	66.7	16.7	83.3
Kalpa Block	**20.47**	**72.29**	**67.29**	**75.89**	**84.34**	**74.70**	**25.30**	**21.66**	**78.34**	**37.33**	**62.67**
Marginal	29.03	54.83	70.46	54.83	87.1	67.75	32.25	16.1	83.9	41.9	58.1
Small	16.2	83.8	70.3	89.2	78.4	81.1	18.9	21.6	78.4	16.2	83.8
Medium	13.3	80	53.3	86.6	93.3	73.3	26.7	33.3	66.7	80	20
Nichar Block	**55.41**	**87.80**	**70.27**	**83.51**	**83.78**	**79.69**	**20.45**	**14.86**	**85.14**	**40.54**	**59.46**
Marginal	13.04	95.65	78.26	78.26	95.65	82.60	17.40	13	87	65.2	34.8
Small	87.5	85	70	82	85	77.5	22.5	12.5	87.5	25	75
Medium	27.3	81.6	54.54	100	54.54	81.6	19.4	27.3	72.7	45.5	54.5
All Blocks	**33.32**	**76.88**	**70.59**	**73.32**	**84.00**	**79.99**	**20.06**	**17.32**	**82.68**	**37.33**	**62.67**
Marginal	25.91	70.36	78.81	55.55	85.18	81.48	18.52	12.32	87.68	46.89	53.11
Small	42.85	80.36	69.64	82.12	84.83	80.36	19.64	16.96	83.04	25.00	75.00
Medium	18.74	81.17	53.11	87.47	78.10	74.92	25.43	31.24	68.76	56.27	43.73

Abbreviations used: Mod. = Modernisation; Hort. = Horticulture; F/S = Farming System.

Table 6.9. Initial cost of apple plantation of sample households

(Rs./Plants)

Block/farm size	Various costs				
	Digging of pits	Filling of pits	Cost of plant	Other exp. like watering, FYM and fencing etc.	Total expen-diture
Pooh Block	**12.84**	**5.80**	**13.79**	**5.71**	**38.14**
Marginal	13.10	5.81	12.92	6.08	37.91
Small	12.62	5.82	14.28	5.40	38.12
Medium	13	5.66	14.83	5.83	39.32
Kalpa Block	**12.48**	**4.48**	**12.61**	**5.84**	**35.36**
Marginal	11.80	4.80	11.64	5.23	33.47
Small	12.16	4.10	12.62	6.41	35.39
Medium	14.66	4.73	14.60	5.7	39.19
Nichar Block	**12.29**	**4.01**	**11.90**	**5.11**	**33.32**
Marginal	11.21	3.86	12.04	7.87	34.98
Small	12.7	4.12	11.65	4.02	32.5
Medium	13.09	3.90	12.54	3.27	32.8
All Blocks	**12.53**	**4.72**	**12.73**	**5.56**	**35.53**
Marginal	12.07	4.87	12.18	6.26	35.38
Small	12.50	4.64	12.79	5.24	35.21
Medium	13.81	4.62	13.94	4.89	37.02

Table 6.10. Average expenditure of apple per box upto various distant markets

(Box/Rs.)

Block/ farm size	Marketing operation and expenditure								
	Packing	Local transport	Loading + unloading	Freight upto market	Octroi install tax	State toll tax	Commission in market	Empty box exp.	Total exp.
Pooh Block	**12.55**	**5.16**	**1.65**	**40.80**	**2.57**	**3.48**	**32.37**	**33.57**	**132.14**
Marginal	12.00	3.70	1.81	40.50	2.77	2.80	27.33	29.33	103.36
Small	12.75	6.4	1.30	41.97	2.68	4	34.80	36.51	140.41
Medium	13.83	4.5	3	35.33	1	3.5	40.83	34.16	136.15
Kalpa Block	**12.15**	**3.37**	**2.15**	**29.41**	**2.08**	**3.13**	**32.77**	**34.62**	**119.69**
Marginal	12.48	2.44	2.19	26.61	1.61	2.34	26.29	36.35	110.31
Small	11.62	4.08	2.02	32.48	2.91	3.84	35	33.70	125.65
Medium	12.8	3.53	2.4	27.63	1	3	40.66	33.33	124.35
Nichar Block	**11.71**	**4.83**	**2.77**	**28.29**	**2.48**	**3.30**	**33.07**	**23.44**	**109.90**
Marginal	11.65	3.86	2.50	33.08	2	3.15	31.69	3.13	118.06
Small	11.57	5.27	2.87	26.6	2.92	3.45	32.47	32.72	117.87
Medium	12.36	5.27	3	24.45	1.90	3.04	38.14	32.18	120.34
All Blocks	**12.13**	**4.39**	**2.21**	**32.49**	**2.36**	**3.29**	**32.75**	**30.63**	**120.23**
Marginal	12.08	3.26	2.15	33.08	2.11	2.72	28.17	24.68	108.25
Small	11.96	5.23	2.10	33.35	2.84	3.75	34.03	34.23	127.48
Medium	12.84	4.31	2.72	27.98	1.31	3.11	39.83	33.09	125.18

Table 6.11. Sources of plant protection material and farm machines & equipments of sampe farms

(Multiple response/%)

Block/farm size	Plant protection				Machines/Equipments		
	Agri. Deptt.	Horti. Deptt.	Block	Market	Horti. Deptt.	Block	Market
Pooh Block	**4.17**	**98.57**	**29.09**	**27.50**	**61.76**	**41.14**	**16.17**
Marginal	3.70	100	29.62	48.14	3.70	88.8	14.8
Small	-	100	25	-	100	-	17.14
Medium	30.6	83.8	50.6	95	100	66.6	16.66
Kalpa Block	**23.48**	**92.77**	**59.04**	**49.38**	**96.39**	**14.18**	**15.65**
Marginal	-	100	-	43.38	100	-	-
Small	32.4	100	100	32.4	100	31.8	29.7
Medium	50	60	80	93.33	80	-	13.33
Nichar Block	**8.84**	**96.07**	**58.11**	**22.97**	**94.59**	**0.00**	**46.01**
Marginal	-	100	100	-	100	-	95.9
Small	7.5	97.5	27.5	20	95	-	27.5
Medium	32.2	82.62	81.8	81.8	81.81	-	9.01
All Block	**12.83**	**95.61**	**49.68**	**34.08**	**85.33**	**17.66**	**25.79**
Marginal	1.23	100.00	38.27	34.56	67.90	29.60	32.16
Small	13.38	99.11	50.67	17.85	98.21	10.51	24.99
Medium	40.24	72.24	75.11	89.68	84.37	12.49	12.47

Table 6.12. Distribution and value of FYM between horticulture and agriculture

(Qtls./Rs.)

Block/farm size	Total qty.	Horticulture		Agriculture		Total value
		Qty.	Value	Qty.	Value	of FYM
Pooh Block	**80.03**	**41.33**	**2479.81**	**38.70**	**2321.41**	**4801.22**
Marginal	89.33	42.15	2529	47.18	2831	5360
Small	65.43	34.50	2070	30.93	1855	3925
Medium	123.3	77.48	4649	45.82	2749	7398
Kalpa Block	**62.74**	**31.67**	**1900.10**	**31.07**	**1863.11**	**3763.20**
Marginal	56.48	20.01	1201	36.47	2188	3389
Small	69.23	44.86	2691	24.37	1460	4151
Medium	59.66	23.23	1394	36.43	2186	3580
Nichar Block	**64.05**	**32.36**	**1935.55**	**31.80**	**1907.69**	**3843.24**
Marginal	59.34	20.86	1252	38.48	2308	3560
Small	65.13	36.48	2183	28.75	1725	3908
Medium	70	41.09	2465	28.91	1735	4200
All Blocks	**68.40**	**34.78**	**2086.98**	**33.61**	**2016.28**	**4103.24**
Marginal	68.24	27.63	1658.15	40.61	2436.41	4054.56
Small	66.58	38.59	2315.51	27.98	1678.08	3993.59
Medium	75.15	39.54	2372.47	35.61	2136.53	4509.00

Table 6.13. Consumption of chemical fertilisers of sample households

(Qtls./Farm/Year)

Block/farm size	Types of fertilizer			
	Cane/Urea	Potash	Super phosphate	Total
Pooh Block	**0.45**	**0.37**	**0.42**	**1.23**
Marginal	.34	.32	.48	1.14
Small	.52	.35	.32	1.19
Medium	.50	.66	.76	1.92
Kalpa Block	**1.00**	**0.47**	**0.63**	**2.11**
Marginal	.31	.27	.30	.88
Small	1.56	.35	.78	2.69
Medium	1.07	1.20	.93	3.20
Nichar Block	**0.82**	**0.35**	**0.34**	**1.51**
Marginal	.48	.26	.43	1.17
Small	.95	.30	.28	1.53
Medium	1.09	.72	.36	2.17
All Blocks	**0.78**	**0.40**	**0.47**	**1.65**
Marginal	0.37	0.28	0.40	1.05
Small	1.02	0.33	0.46	1.81
Medium	0.78	0.40	0.47	1.65

Table 6.14. Various methods adopted for marketing by sample households

(Multiple response/%)

Block/	Method of marketing					Why you have adopted this sytem				
farm size	V.M.	FR agent	C.A.	Self	H.A.	M.N.	L.T.R.	L.T.	P.R.	P.H.R.
Pooh Block	**76.47**	**82.00**	**70.59**	**24.62**	**29.40**	**77.82**	**64.48**	**67.16**	**49.26**	**24.79**
Marginal	96.3	88	62.96	25	29.62	96	55	66	25	44
Small	60	80	77.14	25.7	28.5	77.14	68.57	71	65	2.8
Medium	83.33	66.66	66.66	16.66	33.66	-	83.33	50	66.66	66.66
Kalpa Block	**83.01**	**76.97**	**72.29**	**50.60**	**31.32**	**69.87**	**69.84**	**66.23**	**59.00**	**49.24**
Marginal	100	93.5	58.06	35.48	29.03	96.77	61.2	70.9	35.48	9.35
Small	70	70	81.08	72.97	35.13	64.86	75.66	62.16	83.7	75.6
Medium	80	60	80	26.66	26.66	26.66	73.33	66.64	46.66	66.66
Nichar Block	**72.97**	**68.95**	**70.27**	**44.30**	**26.92**	**78.38**	**79.73**	**71.35**	**63.50**	**43.24**
Marginal	69.56	60.96	73.9	43.4	34.78	100	82.6	82.6	95.6	56.52
Small	72.5	75	65	47	22	75	75	67	50	32.5
Medium	81.81	63.63	81.81	36.36	28.37	45.45	90.90	63.63	45.45	54.54
All Block	**77.73**	**75.85**	**71.11**	**40.68**	**29.29**	**75.07**	**71.47**	**68.20**	**57.54**	**39.88**
Marginal	90.12	82.43	64.19	34.24	30.86	97.43	65.21	72.59	49.06	34.29
Small	67.77	74.91	74.11	48.92	28.37	72.32	73.21	66.65	65.82	37.46
Medium	81.25	62.50	78.12	28.12	28.56	28.12	81.24	62.49	49.99	62.49

Abbrivation used: V.M. = Village merchant, FRA = Forwarding agent, CA = Commission agent, Wh.s. = Whole seller, H.A. = Harvest agent, M.N. = Money needs, L.T.R. = Lack of transport, L.T. = Less trouble, P.R. = Price Risk, PHR = Physical risk.

Table 6.15. Net income from all fruit crops of sample households

(Rs./Farm/Year)

Block/ farm size	Total fruits value							Total income	Total exp.	Deficit/ surplus
	Apple	Almond	Apricot	Walnet	Plum	Pear	Grapes+ bemmi			
Pooh Block	**38115.82**	**5065.59**	**397.01**	**236.00**	**0.00**	**543.90**	**329.13**	**44687.45**	**16228.73**	**28458.73**
Marginal	24730	2600	355.6	–	–	444	385.5	28515.1	10863.3	17651.8
Small	44560	5820	314.28	342.8	–	428.5	276.8	51742.38	16949.8	34792.58
Medium	60761	11760	1066.0	675.0	–	1666.6	380.7	76309.3	36166.9	40142.4
Kalpa Block	**47146.51**	**1726.38**	**149.40**	**417.14**	**0.00**	**421.81**	**438.53**	**50299.76**	**16816.67**	**33483.09**
Marginal	28550	1569.03	96.77	258.85	–	451.6	402.5	31328.75	8919.4	22409.35
Small	51030	924.32	162.16	340.5	–	297.6	386.70	53141.28	18232.9	34908.38
Medium	76000	4030	226.67	933.3	–	666.61	640.85	82497.43	29644.3	52853.13
Nichar Block	**34179.05**	**1730.52**	**0.00**	**658.78**	**1047.30**	**467.18**	**125.42**	**39140.56**	**14329.99**	**24810.57**
Marginal	18560	1323.90	–	782.6	1086.96	652	150	25555.46	8673.7	16881.76
Small	35600	1530	–	600	900	387.5	120.80	39138.3	14752.5	24385.8
Medium	61670	3309.9	–	613.6	1500	370.50	90.80	67553.99	24620.4	42933.59
All Blocks	**40152.38**	**2736.93**	**175.10**	**441.87**	**344.44**	**473.63**	**302.49**	**44933.46**	**15821.14**	**29112.32**
Marginal	24440.00	1843.08	155.57	321.29	308.64	505.97	325.14	28751.54	9497.60	19253.94
Small	43497.41	2670.53	151.78	433.90	321.43	370.61	257.39	47703.06	16588.93	31114.13
Medium	68216.75	5231.84	306.13	774.97	515.63	752.32	402.99	76200.35	29140.32	47060.03

Table 6.16. Information of migratory animals during winter season

(Months/%)

Block/farm size	Season	Name of migratory animals (%)					Look after (%)		
	Winter (months)	Goat	Sheep	Horse	Donkey	Mule	Self	Hired	Contract basis
Pooh Block	**3.89**	**49.99**	**58.82**	**7.35**	**51.47**	**24.99**	**72.05**	**6.43**	**21.52**
Marginal	3.29	44.44	59.25	7.40	48.14	18.51	64.17	2.31	33.52
Small	4.54	51.42	57.14	5.71	45.71	34.28	78.57	9.82	11.61
Medium	2.83	66.67	66.66	16.66	100	0	69.44	5.21	25.35
Kalpa Block	**3.97**	**31.32**	**34.93**	**1.20**	**21.68**	**59.03**	**69.28**	**17.32**	**13.4**
Marginal	3.9	41.93	41.93	0	29.03	0	75.27	14.11	10.62
Small	4.1	18.91	18.91	2.7	0	91.89	65.31	23.65	11.04
Medium	3.8	40.0	60	0	60	100	66.67	8.33	25
Nichar Block	**3.97**	**31.32**	**34.93**	**1.20**	**21.69**	**59.04**	**69.28**	**17.32**	**13.41**
Marginal	6.52	56.52	56.52	0	86.95	4.34	83.3	1.36	15.34
Small	4.2	50.0	50.0	0	47.05	7.50	56.25	12.5	31.25
Medium	3.8	9.09	9.09	9.09	18.18	0	60.6	5.68	33.72
All Blocks	**4.24**	**41.77**	**45.77**	**3.11**	**41.77**	**31.11**	**68.81**	**10.97**	**20.22**
Marginal	4.44	46.91	51.85	2.46	51.85	7.40	74.07	6.56	19.37
Small	4.27	40.17	41.96	2.67	31.25	43.75	66.22	15.34	18.44
Medium	3.61	34.37	43.75	6.25	53.12	46.87	65.1	6.83	28.07

Table 6.17. Information of migratory livestock during the year including local kandas

(Months/%)

Block/farm size	Total time (months)			Look after (%)			Charges by forest for local villagers (%)	
	Rainy	Winter	Summer	Hired	Contract basis	Self/open	No	Yes
Pooh Block	**3.58**	**3.89**	**3.01**	**20.58**	**16.52**	**62.90**	**0.00**	**88.23**
Marginal	3.07	3.29	3.22	7.40	28.52	64.08	0	74.07
Small	4.05	4.54	2.77	31.42	6.61	61.97	0	97.14
Medium	3.16	2.83	3.5	16.66	20.35	62.99	0	100
Kalpa Block	**4.01**	**3.97**	**3.39**	**55.42**	**8.41**	**34.97**	**0.00**	**93.98**
Marginal	4.1	3.9	2.9	45.16	5.62	49.22	0	83.87
Small	3.9	4.1	3.6	75.67	6.04	18.29	0	100
Medium	4.1	3.8	3.9	26.66	20	46.66	0	100
Nichal Block	**3.25**	**4.86**	**3.72**	**25.67**	**21.67**	**52.65**	**0.00**	**89.19**
Marginal	2.43	6.52	3.30	4.34	10.34	85.32	0	73.91
Small	3.6	4.2	3.8	40	26.25	33.75	0	95
Medium	3.7	3.8	4.3	18.18	28.72	53.1	0	100
All Blocks	**3.63**	**4.24**	**3.39**	**35.11**	**15.22**	**49.23**	**0.00**	**90.67**
Marginal	3.28	4.44	3.12	20.98	14.37	64.65	0	77.77
Small	3.96	4.27	3.41	49.10	13.44	37.46	0	97.32
Medium	3.78	3.26	3.96	21.87	23.7	55.06	0	100

Table 6.18. Gross income from sheep (per household) of sample households

(Rs./Nos./Farm/Annum)

Block/farm size	Sale in market		For domestic use		Wool		Total
	No.	Value	No.	Value	(Kg.)	Value	income
Pooh Block	**2.82**	**4365.76**	**0.72**	**1110.53**	**20.21**	**1819.11**	**7295.39**
Marginal	3.14	4867	0.77	1193.5	21.56	1940.63	8001.13
Small	2.45	3797.5	0.57	883.5	15.5	1395	6076
Medium	3.50	5425	1.33	2061.5	41.63	3746.20	11232.7
Kalpa Block	**1.49**	**2310.81**	**0.28**	**439.04**	**6.47**	**582.64**	**3332.49**
Marginal	1.54	2387	0.35	542.5	6.45	580.50	3510
Small	1.0	1550	0.18	279	6.25	562.50	2391.5
Medium	2.6	4030	0.40	620	7.08	636.75	5286.75
Nichar Block	**2.20**	**3406.23**	**0.46**	**714.47**	**12.77**	**1149.12**	**5269.81**
Marginal	2.43	3766.50	0.65	1007.50	13.36	1202.63	5976.63
Small	2.22	3441	0.38	589	12.81	1153.13	5183.13
Medium	1.63	2526.5	0.36	558	11.36	1022.63	4107.13
All Blocks	**2.12**	**3292.13**	**0.47**	**732.56**	**12.70**	**1142.64**	**5167.33**
Marginal	2.33	3605.38	0.58	891.54	13.45	1210.53	5707.44
Small	1.89	2927.70	0.37	578.62	11.48	1033.60	4539.92
Medium	2.44	3774.73	0.56	868.97	15.03	1352.42	5996.12

Table 6.19. Gross income from goats (per household) of sample households

(Rs./Nos./Farm/Annum)

Block/farm size	Sale in market		For domestic use		Wool		Total
	No.	Value	No.	Value	Kg.	Value	income
Pooh Block	**3.48**	**8881.13**	**0.88**	**2233.50**	**50.91**	**3564.01**	**14678.64**
Marginal	3.85	9817.5	1.03	2626.5	53.78	3765.28	16209.28
Small	3.14	8007	0.65	1657.5	39.8	2786	12450.5
Medium	3.83	9766.5	1.50	3825	102.80	7196.74	20788.24
Kalpa Block	**1.51**	**4042.51**	**0.31**	**787.12**	**14.69**	**1027.96**	**5857.60**
Marginal	1.63	4666.5	0.45	1147.5	15.26	1068.4	6882.4
Small	1.35	3442.5	0.21	535.5	14.51	1015.50	4993.5
Medium	1.66	4233	0.26	663	13.93	975.10	5871.1
Nichar Block	**3.33**	**8501.84**	**0.98**	**2586.83**	**27.16**	**1900.77**	**12989.44**
Marginal	3	7650	0.97	2473.5	41.51	2905.80	13029.3
Small	3.97	10123.5	0.90	2295	22.39	1567.10	13985.6
Medium	1.72	4386	0.27	6885	14.47	1012.70	12283.7
All Blocks	**2.71**	**6971.47**	**0.70**	**1816.15**	**29.73**	**2081.47**	**10869.10**
Marginal	2.76	7230.67	0.79	2017.02	35.65	2489.09	11736.78
Small	2.85	7254.98	0.59	1514.52	25.23	1765.78	10535.28
Medium	2.09	5323.13	0.84	2363.44	30.78	2154.58	9841.15

Table 6.20. Gross return from sheep and goats (per household) of sample farm households

(Rs./Farm/Annum)

Block/farm size	FYM of sheep & goats (value)	Total income from goats (wool meat etc.)	Total income from sheep (wool meat mkt.)	Total income from sheep & goats
Pooh Block	**1200**	**14678.64**	**7295.37**	**23174.01**
Marginal	1340	16209.28	8001.13	25550.41
Small	981	12450.5	6076	19507.50
Medium	1850	20788.24	11232.7	33870.94
Kalpa Block	**941**	**5857.60**	**3332.48**	**10131.08**
Marginal	847	6882.4	3510	11239.40
Small	1038	4993.5	2391.5	8423
Medium	895	5871.1	5286.75	12052.85
Nichar Block	**962**	**12989.44**	**5269.79**	**19221.23**
Marginal	890	13029.3	5976.63	19895.93
Small	977	13985.6	5183.13	20145.78
Medium	1050	12283.7	4107.13	17440.83
All Blocks	**1026**	**10869.10**	**5167.32**	**17062.42**
Marginal	1023	11736.76	5707.33	18467.09
Small	1008	10535.28	4539.91	16083.19
Medium	1127	9841.15	5996.1	16964.25

Table 6.21. Consumption of feed/concentrate fed in all seasons

(Qtls./Farm/Year)

Block/farm size	Winter	Summer	Rainy	Total qty.
Pooh Block	**78**	**66.5**	**67.20**	**211.7**
	(1.14)	**(.97)**	**(.98)**	**(3.11)**
Marginal	31	18.05	15.20	64.25
	(1.14)	(0.68)	(0.56)	(2.40)
Small	38	36	40	114
	(1.08)	(1.14)	(3.26)	(3.26)
Medium	9	12	12	33
	(1.5)	(2)	(2)	(5.17)
Kalpa Block	**86**	**77.6**	**79.94**	**243.54**
	(1.04)	**(.93)**	**(.96)**	**(2.93)**
Marginal	31	20.6	22.94	74.54
	(1.0)	(.66)	(.74)	(2.40)
Small	38	35	37	110
	(1.07)	(.95)	(1.0)	(2.97)
Medium	17	22	20	59
	(1.13)	(1.46)	(1.33)	(3.93)
Nichar Block	**69**	**67**	**73**	**209**
	(.93)	**(.91)**	**(.98)**	**(2.82)**
Marginal	21	19	22	62
	(.91)	(.82)	(.95)	(2.70)\
Small	34	36	39	109
	(.85)	(.90)	(.97)	(2.72)
Medium	14	11	12	37
	(1.27)	(1.0)	(1.09)	(3.36)
All Blocks	**233**	**210.1**	**220.14**	**663.2**
	(1.04)	**(.93)**	**(1.03)**	**(2.95)**
Marginal	83	58.1	60.14	201.24
	(1.02)	(.72)	(.74)	(2.48)
Small	110	107	116	333
	(.98)	(.96)	(1.03)	(2.97)
Medium	40	45	44	129
	(1.25)	(1.41)	(1.38)	(4.03)

Note: Figures in parentheses denote the average consumption of concentrate fed (per household).

Table 6.22. Source, quantity and value of concentrates fed of sample households

(Qty./Rs./Farm/Year))

Block/farm size	Purchased from mkt.	Domestic concentrate	Value of market purchased concentrate	Value of demestic concentrate	Total value of concentrate
Pooh Block	**1.03**	**2.06**	**824.35**	**1165.65**	**1990.00**
Marginal	.80	1.60	640	1400	2040
Small	1.09	2.17	872	968	1840
Medium	1.72	3.45	1376	1264	2640
Kalpa Block	**0.98**	**1.95**	**781.49**	**2012.72**	**2794.22**
Marginal	.80	1.60	640	1800	2440
Small	.99	1.98	792	2248	3040
Medium	1.31	2.62	1048	1872	2920
Nichar Block	**0.94**	**1.87**	**750.49**	**2177.08**	**2927.57**
Marginal	.90	1.80	720	2400	3120
Small	.91	1.81	728	1992	2720
Medium	1.12	2.24	896	2384	3280
All Blocks	**0.98**	**1.96**	**784.25**	**1810.77**	**2595.02**
Marginal	0.83	1.66	662.72	1837.04	2499.75
Small	0.99	1.98	794.14	1756.57	2550.71
Medium	1.32	2.65	1057.25	1934.00	2991.25

Table 6.23. Gross income from livestock per household

(Farm/Year/Rs.)

Block/farm size	Total income excluding sheep & goat livestock					Income of sheep + goats	Total earning sheep + goat & other livestock
	Pony/horse mule/donkey	Cows earning	Bullock earning	FYM value	Sub total		
Pooh Block	**5714.12**	**7144.68**	**1973.38**	**3601.09**	**18433.26**	**23174.25**	**41607.28**
Marginal	6240	8424	2230	4020	20914	25550.41	46464.41
Small	4920	5994	1680	2944	15538	19507.50	35045.50
Medium	7980	8100	2530	5549	24159	33870.94	58029.94
Kalpa Block	**3008.67**	**6296.53**	**2674.82**	**2822.39**	**14802.41**	**10130.91**	**24932.49**
Marginal	2100	6426	2220	2542	13288	11239.40	24527.40
Small	3360	5238	29.20	3113	14631	8423	23054
Medium	4020	8640	3010	2685	18355	12052.85	30407.85
Nichar Block	**2028.65**	**5687.51**	**2717.57**	**2882.43**	**13316.16**	**19221.23**	**32537.39**
Marginal	1320	6102	2920	2670	13012	19895.93	32907.93
Small	2400	4860	2540	2931	12731	20145.73	32876.73
Medium	2160	7830	2940	3150	16080	17440.83	33520.83
All Blocks	**3504.00**	**6352.56**	**2476.89**	**3077.48**	**15410.92**	**17062.42**	**32473.35**
Marginal	3258.52	7000.00	2422.10	3071.00	15751.63	18467.76	34218.72
Small	3504.64	5339.25	2396.79	2995.19	14235.87	16083.19	30319.06
Medium	4123.13	8260.31	2895.94	3381.84	18661.22	16964.25	35625.47

Table 6.24. Net return after all costs of livestock of sample households

(Farm/Year/Rs.)

Block/farm size	Grass cereals hay, farm forestry CPRs (1)	Concentrate fed to animal (market + domestic) (2)	Exp. of veterinary + salt + cereals etc. (3)	Look after during migration period (4)	Look after at home (5)	Total exp. (1 to 5) (6)	Total income of livestock (7)	Overall livestock deficit or surplus (8)
Pooh Block	**5367.72**	**2000.00**	**696.25**	**10500.00**	**3580.00**	**22078.97**	**41607.52**	**19528.54**
Marginal	4524.80	2040	715	10800	4100	22179.80	46464.41	24284.61
Small	5667.60	1840	640	9300	3060	20507.60	35045.50	14537.9
Medium	7411.60	2640	940	15300	4500	30791.6	58029.94	27238.34
Kalpa Block	**4781.98**	**2800.00**	**997.83**	**16500.00**	**2380.00**	**27410.90**	**24933.32**	**-2477.58**
Marginal	4489.09	2440	865	13800	2980	24574.09	24527.40	-46.69
Small	4552.40	3040	1090	18300	1740	28722.40	23054	-5668.40
Medium	5933.60	2920	1045	17400	2740	30038.6	30407.85	369.25
Nichar Block	**6124.25**	**2920.00**	**1033.85**	**17409.00**	**3980.00**	**31983.78**	**32537.39**	**533.61**
Marginal	6249.81	3120	1075	18000	5300	33744.81	32907.93	-836.88
Small	6048	2720	970	15900	4900	30538	32876.73	-2338.73
Medium	6139	3280	1180	20100	2860	33559	33520.83	-38.17
All Blocks	**5399.13**	**2595.02**	**918.53**	**14870.67**	**3520.09**	**27303.44**	**32473.35**	**5169.91**
Marginal	5000.95	2499.75	874.63	13992.59	4012.10	26380.02	34218.72	7839.37
Small	5435.04	2550.71	906.52	14630.36	3281.07	26803.70	30319.06	3515.36
Medium	6281.33	2991.25	1071.72	17934.38	3111.25	31389.93	35625.47	4235.54

Table 6.25. Net income from farm forestry of sample households

(Rs./Farm/Year)

Block/farm size	Timber	Fuelwood	Grass & grazing	Misc. uses	Total income	Total exp.	Deficit/ surplus
Pooh Block	**471.18**	**1298.45**	**1139.35**	**218.44**	**3127.42**	**1449.72**	**1677.70**
Marginal	264	810.15	1008	312	2394.15	1477	917.15
Small	576	1582.35	1316	176	3650.35	1358	2292.35
Medium	792	1839.75	700	45	3376.75	1862	1514.75
Kalpa Block	**572.36**	**1350.18**	**1219.52**	**214.87**	**3356.92**	**1371.49**	**1985.43**
Marginal	720	986.7	1463	116	3285.7	1281	2004.7
Small	388.8	1435.5	1246	323.2	3393.5	1379	2014.5
Medium	720	1890.9	651	152	3413.9	1540	1873.9
Nichar Block	**450**	**1297.50**	**1417.22**	**359.84**	**3524.55**	**1414.95**	**2109.61**
Marginal	156	1044.45	1701	816	3717.5	1302	2415.5
Small	624	1313.40	1330	136	3403.4	1484	1919.4
Medium	432	1768.8	1141	220	3561.8	1400	2161.80
All Blocks	**501.54**	**1317.22**	**1260.31**	**263.63**	**3342.70**	**1409.43**	**1933.27**
Marginal	407.85	944.25	1378.91	380.10	3111.11	1352.30	1758.82
Small	531.30	1437.78	1297.88	210.34	3477.30	1409.94	2067.36
Medium	634.50	1839.34	828.63	155.31	3457.78	1552.25	1905.53

Table 6.26. Gross income from major use of CPRs

(Rs./Farm/Annum)

Block/farm size	Value of open grazing (approx.)	Grass collection	Timber wood	Fuel wood	Total income
Pooh Block	**5563.97**	**2858.82**	**2528.74**	**2085.53**	**13037.06**
Marginal	6800	2800	2416	2067.45	14083.45
Small	4450	2880	2850	2055.90	12235.90
Medium	6500	3000	1162	2339.70	13001.7
Kalpa Block	**4830.12**	**1941.93**	**6268.00**	**3129.29**	**16169.34**
Marginal	4950	2240	5566	3399	16155
Small	3850	1620	5374	2704.35	13548.35
Medium	7000	2120	9924	3620.10	22664.10
Nichar Block	**4266.22**	**3265.95**	**4644.11**	**2913.41**	**15089.68**
Marginal	4650	3900	3794	3199.35	15543.35
Small	3500	3200	5260	2481.60	14441.60
Medium	6250	2180	4182	3885.75	16497.75
All Blocks	**4866.44**	**2654.49**	**4603.83**	**2742.84**	**14867.61**
Marginal	5481.48	2898.02	4012.84	2898.46	15290.80
Small	3912.50	2578.04	4544.54	2422.16	13457.23
Medium	6648.44	2305.63	6307.31	3471.34	18732.72

Table 6.27. Gross income from minor use of CPRs

(Rs./Farm/Year)

Block/farm size	Chilgoza	Medicinal herbs	Meat, bones, hides	Mushroom	Flowers/ fruits	Stanch/ fencing etc.	Total
Pooh Block	**1141.54**	**74.99**	**29.37**	**220.57**	**141.18**	**47.06**	**1654.72**
Marginal	814.81	48.14	66.6	133.32	22.20	7.40	1092.47
Small	1089.3	94.28	–	282.84	205.74	68.57	1740.73
Medium	2916.6	83.33	33.2	249.99	300	100.0	3683.12
Kalpa Block	**872.59**	**100.00**	**14.45**	**220.47**	**148.17**	**49.39**	**1405.05**
Marginal	684.10	70.96	32.24	193.53	38.7	12.90	1032.43
Small	843.74	100.0	–	300	267.54	89.18	1600.46
Medium	1333.3	160.0	13.32	79.98	79.98	26.66	1693.24
Nichar Block	**–**	**80.54**	**83.76**	**279.73**	**218.94**	**72.97**	**735.95**
Marginal	–	100.0	34.78	273.9	169.56	56.52	634.76
Small	–	64.0	125.0	277.5	232.55	77.5	776.50
Medium	–	100.0	36.26	300	272.70	90.90	799.86
All Blocks	**666.89**	**86.04**	**41.76**	**239.99**	**169.33**	**56.44**	**1260.45**
Marginal	533.42	71.60	44.41	196.28	70.36	23.45	939.52
Small	619.14	85.36	44.64	286.60	235.73	78.57	1350.04
Medium	1171.85	125.00	24.93	187.49	187.48	62.49	1759.24

Table 6.28. Net returns from CPRs (major and minor use) of sample households

(Rs./Farm/Year)

Type of livestock	Income from CPRs (major use) (1)	Income from CPRs (minor use) (2)	Total income from CPRs (3) (1+2)	Total exp. on CPRs (4)	Surplus or deficit (5) (3-4)
Pooh Block	**13037.06**	**1654.72**	**14691.78**	**7650**	**7041.78**
Marginal	14083.45	1092.87	15176.32	8400	6776.32
Small	12235.90	1740.73	13976.63	6840	7136.63
Medium	13000.70	3683.12	16683.82	9000	7683.82
Kalpa Block	**16169.34**	**1405.07**	**17574.41**	**5825.78**	**11748.63**
Marginal	16155	1032.43	17187.43	6720	10467.43
Small	13548.35	1600.46	15148.81	4860	10288.81
Medium	22664.10	1693.24	24357.34	6360	17997.34
Nichar Block	**15089.68**	**735.95**	**15825.63**	**9797.84**	**6027.79**
Marginal	15543.35	634.76	16178.11	11700	4478.11
Small	14441.60	776.50	15218.10	9600	5618.10
Medium	16497.75	799.86	17297.61	6540	10757.61
All Blocks	**14867.61**	**1260.45**	**16128.06**	**7683.47**	**8444.59**
Marginal	15290.80	939.52	16230.32	8694.07	7536.25
Small	13457.23	1350.02	14807.25	7171.61	7635.64
Medium	18732.63	1759.24	20491.77	6916.88	13574.90

Table 6.29. Market value of different types of land and labour used by sample farms

(Rs. No. &/Farms/%)

Block/farm size	Market value of land (Rs.)			Farmer's own (No.)		Labour used (%)	
	Crop land	Ghansi	Orchard	Shed	Store	Family	Hired
Pooh Block	**25998.53**	**9383.24**	**61825.44**	**0.61**	**0.04**	**57.21**	**42.79**
Marginal	18600	8630	65600	0.62	0.03	65	35
Small	22500	9730	58318	0.57	0.03	55	45
Medium	79700	10750	65300	0.83	0.17	35	65
Kalpa Block	**13618.86**	**6540.48**	**50186.81**	**0.47**	**0.05**	**54.26**	**45.74**
Marginal	14070	7560	45070	.16	-	58.1	41.9
Small	12385	5500	50061.76	.62	.05	54.1	45.9
Medium	15730	7000	61070	.73	.13	46.7	53.3
Nichar Block	**22726.22**	**6816.14**	**50699.84**	**0.48**	**0.07**	**45.96**	**54.04**
Marginal	19600	8600	48600	.304	.13	43.5	56.5
Small	25631	5681	52070	.57	-	50	50
Medium	18700	7214	50108	.54	.18	36.4	63.6
All Blocks	**20355.58**	**7490.28**	**53872.99**	**0.52**	**0.05**	**52.42**	**47.58**
Marginal	17150.25	8211.98	52915.68	0.35	0.05	56.25	43.75
Small	20276.65	6886.52	53359.06	0.59	0.03	52.92	47.08
Medium	28745.31	7776.69	58094.94	0.68	0.15	40.97	59.03

CHAPTER SEVEN

Contribution and Management of Common Property Resources (CPRs)

Common property resources render a variety of services and benefits to the farmers. However, the contribution of common property resources may not always be direct to the farmers. The contribution of CPRs to different components may show its presence only in the overall systems due to the inter-dependence on each other. The significant contribution of these resources are mainly to livestock and the household sector for the rural poor population.

Identification of Common Property Resources in the Study Area

In the study area people have certain customary rights in the state-owned forests, panchayat forests and village common lands. They use these resources for grazing of animals, collection of grass, animal litter, dead wood for fuel and timber for house building. They also collect minor forest products like mushrooms (guchhi), flowers, fruits, medicinal herbs, shrubs, fibres, chilgoza (neoza), meat, bones etc. from government land. Irrigation water is another important CPRs in this area, particularly from the point of view of farming.

Source of Irrigation

Kuhl (gravity irrigation channel) is an important source of irrigation in this dry zone. It contributes significantly to agriculture as well as horticulture component of farming systems. The kuhls derive their water from small rivulets (*nullahs, khads*) fed by melting of snow in the peaks and from underground resources. The water from these rivulets and *khads* is generally tapped at the higher level and carried through open unlined/lined channels and sometimes through PVC/GI pipes to the irrigation points. Some of these irrigation schemes have been executed by Irrigation and Public Health Department (I & P.H.) and Desert Development Project (DDP). Generally these kuhls are collectively maintained and managed by the villagers. However, the distribution of water from these sources at times may not be equitable and may favour the rich and influential sections.

Table 7.1* reveals that kuhl contributes about 70 percent of the water required during summer and rainy seasons and 16 percent during rainy and winter seasons. Nullah is the next important source of irrigation which contributes nearly 5 percent of its water requirements throughout the year. The role of tank and lift irrigation is negligible because of the topographical features of the area.

Water-Driven Flour Mills (Gharats)

Water of the *khads* is commonly used in *gharats* for grinding local millets, wheat, barley, maize, etc. For some people it is a source of their livelihood. It is noticed that the practice of these traditional water-driven flour mills is declining day-by-day. On an average there are 2 to 3 *gharats* per village in the study area.

Major Use of Government/Panchayat Forests

Grazing

The people frequently use forests for grazing and there is no restriction regarding the number of cattle that might be grazed in any forest. Grazing rights of the people are affecting the

*See Tables 7.1 to 7.6 at the end of this chapter.

forests management. Large flocks of sheep and goats graze in the demarcated forests in the late spring and autumn. Sheep and goats are migrated to the alpine pastures during the summer.

The summer grazing upto 1962 did not pose much problems and cattle were grazed in the extensive alpine pastures of western Tibet. But after occupation of Tibet by China in 1962, huge pressure is now on the Indian side. Local people have certain grazing rights, in this regard. The rights and concessions of the local people in forests were admitted and recorded in the forest settlement report completed in 1921.

Building Timber

In the forest settlement, there is a provision of timber to the right-holder for construction of buildings, animal sheds and water channels etc. at concessional rates or free of cost under specific circumstances. No limit on the number of trees to be sanctioned was fixed.

Timber Distribution Rights (TD)

The farmers of Kinnaur depend upon forests for their timber requirements. The result of the analysis has been presented in Table 7.2. The table reveals that in Pooh block each farmer had applied for an average of 1.89 trees and obtained only 0.82, thus showing a deficit of 1.07 trees. In Kinnaur district per farm deficit is 1.03 trees based on average of 5 years. The deficit figure is 1.06 trees in case of Kalpa and 1.0 trees in case of Nichar block respectively. This deficit results in elicit felling of the trees to meet out their requirements.

The farmers are using the wood from forests for various purposes like residential home, farm implement support system, furniture items and other miscellaneous uses. Maximum portion of wood required is extracted from the government land. At overall level, in all cases, CPRs account for nearly 73 percent in residential, 94 percent in replacement of farm implements, 48 percent in furniture items and 97 percent in miscellaneous uses (fencing and staking etc.) respectively (Table 7.3).

Fuelwood

Forests are a great source of fuelwood for villagers and labourers. They extract fuelwood from the forest by lopping and cutting the trees. The fuelwood is generally used by inhabitants and highway labour colonies. Both types of wood i.e. soft and hard is used for burning.

In the study area, which falls in the dry temperate zone, fuelwood is required in a large quantity due to cold climatic conditions almost throughout the year. Table 7.4 shows that on an average farm household of Pooh block consumes 31.12 qtls. of fuelwood per year, out of which CPRs contribute about 20 qtls., in Kalpa the total consumption is 30.70 qtls. out of which CPRs contribute 21 qtls., for Nichar 30.04 qtls. out of which CPRs account for 22 qtls. respectively. It is also seen from the table that dependence on forests is declining due to substitution of other sources of fuel such as LPG, electricity and solar & biogas in the study area (Table 7.5).

Cutting of Torch Wood

The villagers collect the resinous torch wood from the basal portion of the stem of kail, neoza pine and deodar. Now, this trend, however is on the decline due to electrification in the area. But the use of torch wood is still prominent among the poor households in the villages and in labour colonies.

Lopping

Miscellaneous broad leaved tree lopping, specially near habitation, along road sides and in rivers banks where large flocks of sheep and goats are grazed during migration is prevalent in the study area.

Chilgoza (Neoza)

Neoza (seeds of neoza pine) in Kinnaur is of great economic importance and people have got rights to extract these pine seeds. In these forests, the cones appear in June-July and mature in September-October next year, taking about 15-16 months to mature. Resinous cones are collected and stored for 15-45 days to allow them to warp and facilitate extraction of seeds. It is generally noticed that all the cones are extracted from the pine trees and due to this total extraction of cones, the problem of re-germination has become very common. Neoza forests are generally found in Kalpa, Morang, Pooh and Nichar forest ranges.

MINOR FOREST PRODUCTS (FLORA)

Minor forest products like medicinal herbs etc. are collected by the right-holders in the respective revenue circles. Main medicinal herbs like Dhoop, Kuth, Patish & Kala Zira are collected by villagers from the wet and dry area of Kinnaur and exported for sale. The right of cutting grass (another minor products) has also been noticed even in the demarcated and protected forests. People also collect fruits, honey, edible seeds, flowers, roots, leaves, small branches, bark, dyes, wild mushrooms (guchhi) etc. from the forests to meet their households demands and for sale.

Thuni (Thuniara)

The bark of this plant is used as a substitute of tea in Kinnaur. The fruits of this tree are edible but the seeds although having medicinal value, are very poisonous. The wood of this tree is also used for construction purposes due to its water-resistant property.

Birch Tree (Bhojpatra)

The bark of the Birch tree was used for writing purpose in the old days. The old literature written on Birch bark can be found in gompas. Now the practice has been abandoned due to the use of commercial paper. Now the use of this tree has got limited up to the construction of house-roofs for the control of leakage problem.

Reonsh

Reonsh, an important plant of Kinnaur district is found at high altitudes. The stems and branches of the tree are used for making walking-sticks, handles for agricultural tools and plough-shares.

Other Uses of Forests

The trees from the forests are used commonly for staunching, sheep & goat folds, furniture, support systems for fruit plants. Small plants and bushes are used for providing support to the crop of pea, tomato etc. People commonly use local mine stones for house construction.

Some other important common property resources are wells (*bauri*) ponds, lakes, river banks threshing grounds, temples, common paths, etc. in the district. These minor resources have significant impact on the farming systems.

WILD LIFE (FAUNA)

The considerable variation in elevation and climate, which is an essential requirement for rich wild life, is prevalent in this area. Main animals and birds found in this area are musk deer, goral, black bear, panther, snow leopard, brown bear, tibetan sheep, tibetan wolf, tibetan wild ass, porcupine, thar, bharal, ibex, monal, koklas, chakor, black partridge, wood peckers, cheer pheasant, jungle fowl, snow pigeon, snow cock, and tragopan etc. apart from the meat consumption, these wild animals and birds are also generally used for medicinal purposes especially musk deer and black bear. However, the animals and birds are mostly used for meat, skin and bones.

Management of Common Property Resources in the Study Area

The study reveals that state forests, Panchayat forests and village common lands are the major common property resources. The irrigational channels, which are popularly called as kuhl in this area seem to be under great stress. The kuhls are the most important source of irrigation because the area falls in dry temperate zone. During the survey, information regarding various uses of CPRs and different attitude of the respondents towards CPRs management were collected. The results of this analyses are outlined below.

Management of the Kuhls

Kuhls are the channels constructed alongwith the hill slopes. In these kuhls water is received from local *khads* & streams and leads to agricultural and horticultural land. In the study area two blocks i.e. Pooh and Kalpa are nearly 100 percent irrigated whereas some part of Nichar block is also irrigated. Therefore, irrigation is very important in Kinnaur. During this survey, it is observed that generally the length and command area of the kuhl channel decreases as one moves from topmost kuhl of the *khad* to the lowermost kuhl.

The management of these kuhls rests with the farmers. However, some of the kuhls have also been taken over by the Department of Irrigation & Public Health (I. & P.H.) But the overall management and operations is under the local villagers. Details have been given in Table 7.6.

Financial Help for Kuhls

Generally, the repair and maintenance is carried out by local people, but financial help is also provided by the government from time-to-time. Many a times natural calamities completely destroy these channels. In such devastating conditions, special financial help is provided by Department of Agriculture and Irrigation and Public Health Department (I & PH).

Kuhl Administration

We have seen in other areas of H.P. that kuhl management has generally failed. But in Kinnaur there is no compromise about the maintenance and performance of kuhls, because it is the part and parcel of farming system. Kuhl committee is formed yearly, which decides the distribution and take the responsibilities of kuhl maintenance.

It was observed that government is not paying proper attention towards management of the kuhls. In the study area there is also a water scarcity. So, majority of the kuhls are still used jointly. Many times the water of these kuhls is also used for domestic purposes such as for washing the clothes, cleaning utensils and livestock care etc. But the use of kuhls for domestic purpose is now declining because government has provided drinking water almost in every village. Kuhls are also used for *gharats*. The committee formed for kuhl management performs the following operations.

- Cleaning & digging of kuhl
- Retaining wall (*danga*)
- Other repairs
- Water distribution
- Conflict management

It was reported that some farmers do illegal activities on kuhls. They divert the water by breaking retaining walls etc. But these conflicts are generally settled down by the kuhl management committee.

FOREST

Panchayat and government forests are main source of fuelwood, fodder, grazing and timber. Panchayat forests are owned by revenue department and managed by forest department. The ownership and management of the state forests however lie with the state forest department. But the inhabitants have got the customary rights in these forests. These rules and rights are also embodied in the forest settlement records. The rights for the use of forest products are clearly defined for each forest and village falling under particular forest area.

Main Rights

The major forest rights include timber for building purposes, fire-wood for funeral, marriage ceremonies, timber for agricultural implements, grazing, grass cutting, lopping of trees, collection of medicinal herbs, fibres for ropes, flowers & fruits etc. The government is entitled to protect, improve and reproduce the forest growth and sell trees to the traders and non-right-holders only after meeting the requirements of the right-holder. These rights are exercised and regulated by rules framed under section 32 of the Indian Forest Act as notified in Gazette No. 416 dated August 1877 and subsequent amendment of section 55 dated 6th February 1904, and 995 on dated 11th January 1919. But these rights have been overexploited by the beneficiaries. Now the government has curtailed these rights in the new National Forest Policy. Under Forest Conservation Act, 1980, the people of the villages are not allowed to encroach forest land for cultivation and grazing of livestock in restricted areas. Similarly, the right of coal making and small timber for agricultural implements has also been curtailed. But despite of these restrictions, the contribution of CPRs is significant in the study area regarding grazing, collection of grass, animal litter, fuelwood etc.

The villagers contribute very little for sustaining the productivity of the CPRs by leaving behind the animal droppings only, inspite of extracting several benefits. People do not look after these resources by thinking that it is the sole responsibility of the state government. With the increase of population the demand for forest products has immensely increased and thus their management will require efforts of all.

Animal grazing, illegal felling, excessive lopping of green trees and shrubs in forest are identified as the major threats by the forest department. To check these activities, fencing, imposition of fine on defaulters are the major steps. But people's participation is negligible regarding forest maintenance.

The views of the villagers about the participation and efficient management of the forest were also recorded. Majority of the respondents are in favour of dividing of total forest area of the village among the individuals alongwith exclusive rights for grazing/grass collection. Similar views are quoted in case of panchayat forests.

Common Land Management

Village common land ownership is vested with the state revenue department. The utilisation pattern of village common land is existing for several years. Earlier the breaking of land for cultivation was a principal right. Now, these rights have been curtailed under new policies. The village common land is now used for staking of wood, grass, local fairs & festivals, staking of animal litters etc.

The conditions of village common lands and forests are deteriorating and inhabitants are fully aware of this condition, yet, they have not approached any government institution in this regard. It was noticed that management of CPRs is very poor. Moreover, the availability of village common land has declined due to encroachments and allotment. But increasing livestock pressure has aggravated the position further.

It is revealed that the villagers want to control this deterioration by dividing the land equally among themselves. So that they can make the best use of this land by cultivation and through plantation crops. But some respondents favour the handing over of the land to forest department for bringing it under tree cover and grasses.

Conclusions

It can be concluded from the above discussion that inadequate management of CPRs is responsible for the deteriorating condition. This has resulted in considerable decline in the production potential of CPRs in the area. The government is unable to arrest the continuously declining status of CPRs due to inadequate manpower, funds and users' non-cooperation. Alongwith this, there are cultural and political constraints which hinder the implementation of policy programmes. Therefore, the prevailing systems need changes regarding the CPRs utilisation and management to maintain the balance between economic and ecological uses.

Suggestions

The following suggestions would contribute towards the better management of these common property resources in Kinnaur district of H.P.

- The farmers of the study area are dependent on common property resources (CPRs). Efforts should be made towards protection, regeneration and development of their CPRs.
- Encroachment and other illegal activities in the CPRs should be checked.
- An area in a pasture should be earmarked and be closed for production of seeds of desirable grasses.
- Provision for the grasses from the particular enclosures for a particular month in a year should be made.
- Rotation/different grazing systems may be followed to give the advantage of early nutritive growth.
- Mining and quarrying activities should be stopped immediately.
- Timber distribution (T.D.) is generally more than their genuine needs in some case. Therefore, T.D. should be regulated only for genuine needs.
- Due to cold desert region, there is an acute shortage of fuel and timber wood and people depend largely on forests. So, people should be encouraged to use LPG, biogas and solar energy etc.
- Multipurpose tree species should be preferred.
- Various varieties of grasses can also be incorporated in the

degraded forest sites and common lands.

- Benefits of keeping improved animal breeds under stall-feeding should be demonstrated.
- Scientific exploitation of minor forest products should be encouraged to supplement the income of the people.
- Chilgoza forests, being a source of income to the local people should be put under effective management.
- The region is endowed with rich fauna which should be carefully preserved.
- Undesirable growth of weeds and unpalatable grass species should be eradicated manually or by the use of weedicides to obtain high production level.
- Common paths, temples, underground water resources which are used commonly, should be maintained properly.
- People's involvement can be increased by linking the benefits with the management responsibilities under the joint management of CPRs.

For the better management of kuhls, the following suggestions should be followed.

- Irrigation channels must be *pucca* or lined to avoid seepage to the nearby areas prone to landslides.
- Besides tapping more surface water lift irrigation should be made functional.
- Kuhl committee should be strengthened to improve the implementation aspect.
- If a member of kuhl cannot participate in any of kuhl operation, he should deposit the equivalent amount with the committee to hire labour well in time.
- Financial aid from government for repairs and development works may be invested on the direction of kuhl committees.
- A joint account from the contribution of members should be raised for any unforeseen repairs of kuhls.
- Government intervention in kuhl committee should not be biased.
- Irrigation, public health department and other government

agencies should pay more attention towards irrigation facilities.

- Water users' right should be maintained as per the revenue records and water distribution should be maintained as per the traditions.
- Minor repair should be done by the water users under the control of kuhl committee. The repair and maintenance and other departmental works are required to be completed every year.

Table 7.1. Sources of irrigation of sample farm households

(Km./Season/%)

Block/farm size	Kuhl			Nullahs			Tank			Lift		
	Dis-tance	Supply S+R	Supply R+W	Dis-tance	Supply S+R	Supply R+W	Dis-tance	Supply S+R	Supply R+W	Dis-tance	Supply S+R	Supply R+W
Pooh Block	**2.69**	**100**	**10.29**	**1.51**	**2.94**	–	**0.01**	**4.41**	–	**.01**	**1.47**	–
Marginal	1.66	100	22.22	0.22	3.70	–	–	–	–	–	–	–
Small	3.37	100	2.85	–	–	–	–	–	–	–	–	–
Medium	3.33	100	–	16.16	16.66	–	0.16	50	–	0.166	16.66	–
Kalpa Block	**2.53**	**79.52**	**28.91**	**0.12**	**8.43**	**3.61**	–	–	–	–	–	–
Marginal	1.62	100	–	0.32	22.58	9.67	–	–	–	–	–	–
Small	2.91	54.05	37.83	–	–	–	–	–	–	–	–	–
Medium	3.46	100	66.66	–	–	–	–	–	–	–	–	–
Nichar Block	**0.75**	**31.08**	**5.40**	**0.03**	**2.70**	**2.70**	**0.01**	**1.35**	**1.35**	–	–	–
Marginal	0.43	30.43	8.69	0.086	8.69	8.69	–	–	–	–	–	–
Small	0.87	30	5	–	–	–	0.025	2.50	2.50	–	–	–
Medium	1	36.36	–	–	–	–	–	–	–	–	–	–
All Blocks	**1.99**	**69.78**	**15.55**	**0.51**	**4.89**	**2.22**	**0.01**	**1.78**	**0.44**	–	**0.44**	–
Marginal	1.30	80.25	9.87	0.22	12.34	6.17	–	–	–	–	–	–
Small	2.33	59.82	50.17	–	–	–	.01	.89	.89	–	–	–
Medium	2.59	78.12	31.25	3.03	3.12	–	0.03	9.38	–	0.03	3.12	–

Abbreviation used: W = Winter, S = Summer, R = Rainy.

Table 7.2. Information regarding timber distribution rights (applied/obtained) of sample farm households

(%/Nos.)

Block/farm size	Have you applied for T.D. in last five years (% response)		How much obtained avg. in 5 years (No. of trees)	
	Yes	No	Applied	Obtained
Pooh Block	**69.11**	**30.89**	**1.89**	**0.82**
Marginal	77.77	22.23	2.48	1.44
Small	62.85	37.15	1.31	0.25
Medium	66.66	33.34	2.66	1.33
Kalpa Block	**86.74**	**13.26**	**2.21**	**1.15**
Marginal	83.87	16.13	2.61	1.19
Small	83.78	16.22	1.51	0.91
Medium	100	–	3.13	1.66
Nichar Block	**81.08**	**18.92**	**2.12**	**1.10**
Marginal	86.95	13.05	2.43	1.21
Small	72.5	27.50	1.7	0.9
Medium	100	–	3	1.63
All Blocks	**79.55**	**20.45**	**2.09**	**1.03**
Marginal	82.71	17.29	2.52	1.28
Small	73.21	26.79	1.52	0.70
Medium	93.75	6.25	3.00	1.59

Table 7.3. Use of wood and their sources of sample households

(Multiple response/%)

Block/ farm size	Residential house			Impliments replacement			Misc. uses			Furniture	
	Govt. land (CPRs)	Own land	Market land	Govt. land (CPRs)	Own land	Market	Govt. land (CPRs)	Own land	Market	Govt. land (CPRs)	Market
Pooh Block	**95.59**	**7.35**	**47.05**	**88.23**	**48.52**	**69.11**	**94.11**	**80.88**	**–**	**47.05**	**84.7**
Marginal	50	18.51	29.62	85.18	6.60	48.14	92.59	66.66	–	59.25	74.07
Small	94.28	9.25	68.57	88.57	16.7	97.14	94.28	88.57	–	42.85	100
Medium	83.33	15.25	–	100	10.50	87.10	100	100	–	16.66	100
Kalpa Block	**63.01**	**10.31**	**15.48**	**93.97**	**15.54**	**24.40**	**96.38**	**46.98**	**–**	**54.21**	**73.49**
Marginal	80	7	10	100	12.9	10.50	96.77	22.58	–	48.38	61.29
Small	50	12	15	100	18.64	32.43	94.59	64.86	–	64.86	72.97
Medium	60	13	28	66.66	13.33	33.33	100	53.33	–	40	100
Nichar Block	**82.37**	**16.16**	**13.04**	**98.65**	**12.16**	**28.59**	**100.00**	**39.19**	**–**	**43.24**	**66.21**
Marginal	91.3	12	17.39	100	8.69	34.78	100	21.73	–	39.13	65.21
Small	80.5	17.5	10	100	15	27.9	100	55	–	37.5	62.5
Medium	70.5	20	15	90.9	9.09	18.18	100	18.18	–	72.72	81.81
All Blocks	**73.22**	**13.19**	**24.22**	**93.78**	**13.40**	**41.62**	**96.89**	**54.66**	**–**	**48.44**	**76**
Marginal	73.21	12.26	18.26	95.06	9.60	29.94	96.29	37.03	–	49.38	66.66
Small	74.73	13.10	29.95	96.43	16.73	51.03	96.43	68.75	–	48.21	77.68
Medium	67.98	15.83	18.28	81.24	11.34	38.20	100	50	–	46.87	93.75

Table 7.4. Source of fuelwood in different seasons of sample households

(Qtls./Farm/Annum)

Block/ farm size	Winter (20th October-20th March)				Summer (21st March-20th June)				Rainy (21st June-19th October)				Total con-sump-tion
	Own land	Govt. land (CPR)	Farm forestry	Market	Own land	Govt. land (CPR)	Farm forestry	Market	Own land	Govt. land (CPR)	Farm forestry	Market	
Pooh Block	**3.67**	**10.18**	**0.47**	**1.11**	**1.92**	**4.91**	**0.19**	**0.01**	**2.32**	**4.94**	**1.60**	**0.00**	**31.12**
Marginal	2.51	8.83	1.18	1.59	1.48	3.92	0.48	0.03	0.92	2.44	4.03	——	26.91
Small	4.48	10.88	——	0.14	2.11	5.37	——	——	3	6.34	——	——	32.32
Medium	4.16	12.16	——	4.66	2.83	6.66	——	——	4.6	8	——	——	43.07
Kalpa Block	**3.49**	**9.45**	**0.00**	**1.68**	**2.09**	**5.00**	**0.00**	**0.00**	**2.42**	**6.54**	**0.02**	**0.00**	**30.70**
Marginal	3	10.06	——	2.25	1.33	3.96	——	——	1.16	5.93	0.06	——	27.75
Small	3.54	8.4	——	1.08	2.35	5.18	——	——	2.81	6	——	——	29.36
Medium	4.4	10.8	——	2	3	6.73	——	——	4.06	9.13	——	——	40.12
Nichar Block	**3.20**	**10.12**	**0.00**	**0.00**	**2.05**	**5.37**	**0.00**	**0.00**	**2.62**	**6.68**	**0.00**	**0.00**	**30.04**
Marginal	3.13	11.86	——	——	1.34	5.3	——	——	1.86	6.21	——	——	29.70
Small	2.97	8.52	——	——	2.17	4.9	——	——	2.82	6.22	——	——	27.6
Medium	4.18	12.27	——	——	3.09	7.22	——	——	3.45	9.36	——	——	39.62
All Blocks	**3.45**	**9.89**	**0.14**	**0.96**	**2.02**	**5.10**	**0.06**	**0.00**	**2.45**	**6.10**	**0.49**	**0.00**	**30.61**
Marginal	2.87	10.16	0.39	1.39	1.38	4.33	0.16	0.01	1.28	4.85	1.37	0.00	28.02
Small	3.63	9.22	0.00	0.40	2.21	5.14	0.00	0.00	2.87	6.18	0.00	0.00	29.66
Medium	4.28	11.56	0.00	1.81	3.00	6.89	0.00	0.00	3.95	9.00	0.00	0.00	40.50

Table 7.5. Types of fuel consumption of sample farm households

(Rs./Farm/Annum)

Block/farm size	Energy sources					
	Fuelwood (Qtls.)	coal (Qtls.)	Agri. waste (Qts.)	K. oil (Ltrs.)	Electricity (Rs.)	L.P.G. (Cylenders)
Pooh Block	**31.12**	**0.60**	**4.95**	**6.97**	**733.83**	**4.55**
Marginal	26.91	0.81	2.92	3.94	570.41	4.33
Small	32.32	0.37	5.17	8.17	785.7	4.48
Medium	43.07	1	12.83	13.64	1166.66	6
Kalpa Block	**30.70**	**0.49**	**5.89**	**2.28**	**693.97**	**4.48**
Marginal	27.75	0.48	4.22	3.51	364.5	3
Small	29.36	0.48	5.92	1.75	764.86	4.64
Medium	40.12	0.53	9.26	1.06	1200	7.13
Nichar Block	**30.04**	**0.27**	**6.58**	**7.74**	**1093.24**	**4.37**
Marginal	29.7	0.17	6.39	10.78	765.21	2.52
Small	27.6	0.2	5.97	2	1287.5	4.62
Medium	39.62	0.72	9.18	22.27	1072.72	7.36
All Blocks	**30.61**	**0.45**	**5.83**	**5.50**	**837.33**	**4.47**
Marginal	28.02	0.50	4.40	5.72	546.92	3.31
Small	29.66	0.35	5.70	3.85	958.03	4.58
Medium	40.50	0.68	9.90	10.71	1150.00	7.00

Abbreviations used: Agri. = Agriculture; K. Oil = Kerosene oil; L.P.G. = Liquified Petroleum Gas.

Table 7.6. Management and maintenance of irrigation system of sample farms

(Multiple response/%)

Block/farms size	Turn of irrigation		Maintenance		How many times of turn		
	Vill. committee	Panchayat	All villagers	Panchayat/ I. & P.H.	Once a week	Fort-nightly	Once a month
Pooh Block	**100**	–	**98.61**	**7.35**	**45.59**	**41.18**	**13.23**
Marginal	100	–	96.49	3.70	37.04	62.96	–
Small	100	–	100	–	48.57	22.73	25.7
Medium	100	–	100	66.66	66.66	33.34	–
Kalpa Block	**100**	–	**100**	**4.82**	**39.76**	**44.58**	**15.66**
Marginal	100	–	100	–	51.63	32.25	16.12
Small	100	–	100	–	21.62	56.76	21.62
Medium	100	–	100	26.66	60	40	–
Nichar Block	**40.54**	–	**39.19**	**2.70**	**25.67**	**13.51**	–
Marginal	30.43	–	30.43	–	26.08	4.34	–
Small	30	–	27.50	5	20	7.5	–
Medium	100	–	100	–	45.45	54.54	–
All Blocks	**80.44**	–	**79.55**	**4.90**	**36.90**	**33.30**	**9.77**
Marginal	80.24	–	79.01	1.23	39.50	34.56	6.17
Small	75	–	74.10	1.78	29.46	29.46	15.17
Medium	100	–	100	25	56.25	43.75	–

CHAPTER EIGHT

Quantification of Linkages Between Different Components of Farming Systems

Introduction

It is revealed that crop husbandry, livestock husbandry and forestry constitute the main and closely integrated components of farming systems. A change in any one of the components of farming system significantly affects the others. In the production process each sub-system contributes in term of employment creation as well as income generation. All these effects and inter-relations in the form of linkages are of great consequence. A quantitative analysis of any one component of the farm economy may be incomplete without taking into consideration the relationship existing among different activities at the same time.

Over centuries, the forests and other common property resources have formed an important part of the tribal economy of Himachal Pradesh. A strong relationship has always existed between forestry and life, culture and economy of the great multitude of rural populace. Tribals fulfil major requirements of fuelwood, timber, fodder, litter, grazing, fibres and fruits etc.; either from own private forest land or from public forests or from both. In fact, all the sub-systems of tribal agriculture are closely inter-connected with farm land, forestry and community resources like forests and community land. However, the nature and degree of dependence varies depending upon the nature of local economy, income status of the people and agro-climatic conditions.

The average tribal farmer is too poor to afford stall-feeding of animals. The usual practice is to send livestock for grazing in forests, cultivable wasteland and in the agriculture fields (after the crops are harvested). Grazing in the forests has become a long standing tradition and continued practice of grazing has automatically vested in the graziers' right which is almost legal of leading their animals into forests. Grazing thus helps in livestock production in the region. In addition to grazing there is a massive supply of cut-fodder from forest areas in almost all seasons and particularly in times of scarcity. Government forests in the tribal region provide timber and poles for livestock sheds and housing, farm implements & tools and furniture etc. Wood for making various farm implements & tools and packing cases are renewed regularly to sustain the hill farming practices. Among the most important products derived from the forests are fodder for stall-feeding. Forest fodder is particularly important for livestock maintenance during the dry season when it is characteristically in short supply. Declining availability of fodder from forests, agricultural residues, natural pastures is limiting the stall-feeding of livestock at night time only. Free range grazing and browsing constitute the other part of the fodder consumption in tribal region of Himachal Pradesh.

Another very important contribution of forests to tribal farming system is the use of plant biomass as bedding material for animals. Forests biomass when mixed with animal excreta yields the organic compost which forms the principal source of soil nutrients for tribal agricultural land. The yields of crops are likely to decrease when fodder and leaf litter are no longer available in sufficient quantities from forested lands. So, there is a net transfer of fertility from forests to crop land via livestock component of the system. The remarkably high dependence of crop lands on organic manure from animal dung and forest biomass clearly shows how the land crop sub-system, animal husbandry and forestry are inter-connected with each other in the tribal farming systems.

Government forests and farm land forestry constitute the important source of fuelwood supply to domestic household sector. Besides, rural people also use forest products for income generation alongwith direct domestic consumption. Many farmers augment their cash income and utilise their spare time by engaging in various home-based small scale industries and trades dependent on forest products. Most of these activities consume forest products as raw material, others require fuelwood as the source of energy for primary and secondary processing activities, while some depend on forest for both raw material and energy supply. Thus, the common property resources (CPRs)-farming system inter-relationship is crucial to the existence and viability of tribal farmers. This chapter highlights the linkages among the different components of the tribal farming system. Though some relations are of qualitative nature yet, an attempt has been made to quantify these relationships between the important sectors of farm economy in the form of interactions with the help of input-output analysis.

Inter-dependence of Different Components of Farming Systems in Pooh Block

Systems Linkages on Marginal Farms

Per household quantitative relationship between various activities of farming system on marginal farms of Pooh block is presented in Table 8.1.* Livestock output worth Rs. 46,464 per annum requires an input of agricultural crops worth Rs. 2117. The contribution of fodder from farm forestry and CPRs comes to Rs. 1028 and Rs. 2847 respectively. The labour input value is Rs. 14,900 and purchased inputs of Rs. 1288. The input-output coefficients of these inputs are Re. 0.05, Re. 0.02, Re. 0.06, Re. 0.32 and Re. 0.03 per rupee of output respectively.

---

*See Tables 8.1 to 8.16 at the end of this chapter.

In the total annual output of fruit crops, the annual contribution of livestock, farm forestry, CPRs, labour and purchased inputs are estimated at Rs. 4609, Rs. 47, Rs. 186, Rs. 4678 and Rs. 1343 respectively. The corresponding values of input-output coefficients work out to be Re. 0.16, Re. 0.002, Re. 0.007, Re. 0.16 and Re. 0.05 per rupee of output.

In total annual agriculture production of Rs. 9053, marginal farmers make use of livestock inputs worth Rs. 4221; own seed worth Rs. 352; farm forestry inputs worth Rs. 30; CPRs inputs worth Rs. 60; labour input worth Rs. 1477 and market-oriented inputs worth Rs. 250. The livestock input coefficient works out to Re. 0.47, seed input coefficient Re. 0.04; farm forestry input coefficient Re. 0.003; CPRs-based input coefficient Re. 0.007; labour input coefficient Re. 0.16 and purchased input coefficient Re. 0.03 in agriculture productions.

In terms of absolute figures, farmland forestry and CPRs depend on labour input to the extent of Rs. 1408 and Rs. 8400 respectively. The total annual value of inputs in respect of livestock, fruit crops, agriculture, farm forestry and CPRs is estimated at Rs. 22,180, Rs. 10,863, Rs. 6390, Rs. 1477 and Rs. 8400 respectively.

Systems Linkages on Small Farms

The quantification of linkages between various sub-systems in respect of small farms of Pooh block is worked out and summarised in Table 8.2. It can be seen from the table that annual livestock output worth Rs. 35,046 requires an input of agriculture worth Rs. 2440; farm land forestry produced fodder inputs worth Rs. 1316; public land produced fodder inputs worth Rs. 2880; labour input worth Rs. 12,359 and purchased inputs valuing Rs. 1375. The input values per rupee of livestock output in respect of agriculture, farm forestry, CPRs works out to Re. 0.07, Re. 0.04 and Re. 0.09 respectively.

In horticultural crops, the annual contribution of livestock, farm forestry, CPRs, labour and purchased inputs is estimated at Rs. 3710, Rs. 26, Rs. 234, Rs. 11,149 and Rs. 1831 respectively. In this sector, each rupee worth of output requires input worth Re. 0.07 from livestock, Re. 0.001 from farm forestry, Re. 0.005 from CPRs, Re. 0.22 from labour inputs and Re. 0.04 from value of purchased inputs.

In agriculture production, the annual contribution of livestock, seed input of its own, forestry, CPRs, labour and purchased inputs is estimated at Rs. 4544, Rs. 762, Rs. 120, Rs. 106, Rs. 8059 and Rs. 503 respectively. In this sector the value of input-output coefficients has been estimated to be Re. 0.23, Re. 0.04, Re. 0.006, Re. 0.005, Re. 0.41 and Re. 0.03 in respect of livestock, agriculture, farm forestry, CPRs, labour and purchased inputs respectively.

The contribution of farm forestry in its own production is Rs. 40. Farm forestry contributes in agriculture for making tools/handles of various implements used for cutting grass, fuelwood and timber. However, farm forestry and CPRs are very important users of labour input in which large quantities of labour is needed for harvesting operations. In terms of absolute figures, farm land forestry and CPRs utilise labour to the tune of Rs. 1292 and Rs. 6840 respectively. The market-oriented inputs in farm forestry is estimated at Rs. 26 annually. The farm forestry and CPRs produce output worth Rs. 3650 and Rs. 13,977 annually.

The total value of inputs in respect of livestock, fruit crops, agriculture crops, farm forestry and CPRs account for Rs. 20,508, Rs. 16,950, Rs. 14,094, Rs. 1358 and Rs. 6840 respectively. The contribution of market-oriented inputs in all systems is constantly increasing. The remaining proportion of total inputs required in these sub-systems of farming system is obtained within the system.

Systems Linkages on Medium Farms

Quantitative relationship between various sub-systems in respect of medium farms in Pooh block is presented in Table 8.3. It is clear from the table that annual livestock output worth Rs. 58,030 requires crop by-products/concentrates worth Rs. 4976, farm forestry inputs worth Rs. 740, CPRs inputs worth Rs. 3160, labour inputs worth Rs. 1980 and market-oriented inputs worth Rs. 2115. In terms of input-output coefficients each rupee of livestock output requires agriculture input of Re. 0.09, farm forestry inputs worth Re. 0.01, CPRs input is Re. 0.05, labour input is Re. 0.34 and purchased inputs is Re. 0.04.

In case of fruit crops, annual output worth Rs. 76,309 of these crops require an input of livestock worth Rs. 7309, farm forestry input worth Rs. 7, CPRs inputs worth Rs. 452, labour and market-oriented inputs work out to Rs. 24,649 and Rs. 3750 respectively. The input coefficients per rupee of output of these crops are estimated at Re. 0.10, Re. 0.0001, Re. 0.006, Re. 0.32 and Re. 0.05 in respect of livestock, farm forestry, CPRs, labour and purchased inputs respectively.

The annual production of agriculture crops worth Rs. 41,275 utilises livestock input of Rs. 6985; seed input of Rs. 1627; farm forestry input of Rs. 40 and CPRs-based inputs of Rs. 99. The value of labour utilised in these crops is Rs.18,448, while purchased inputs value is Rs. 717. The input-output coefficients work out to Re. 0.17, Re. 0.04, Re. 0.001, Re. 0.002, Re. 045 and Re. 0.02 for livestock, seed, farm forestry, CPRs, labour and purchased inputs respectively per rupee of output of agriculture crops.

Vegetable production (other than peas and potato) being a weak activity in this block which utilises negligible amount of inputs has also been included in agriculture sector. Farm land forestry and CPRs make use of labour for doing harvesting operations to the extent of Rs. 1658 and Rs. 9000 respectively.

Systems Linkages in Overall Situation

The dependence pattern of various components with each other, in different farm situations in Pooh block is given in Table 8.4.

On an average, in farming system in Pooh block, input value of agriculture, farm forestry, CPRs, labour, market-oriented inputs required for total annual livestock production worth Rs. 41,607, works out to Rs. 2536, Rs. 1165, Rs. 2948, Rs. 14,016 and Rs. 1414 respectively. In terms of input-output analysis, each rupee of livestock output is using agriculture inputs worth Re. 0.06, labour input worth Re. 0.34 and market-oriented inputs worth Re. 0.03. The fruit crops require livestock inputs worth Rs. 4384, farm forestry Rs. 33, CPRs inputs Rs. 234, labour worth Rs. 9771 and purchased inputs worth Rs.1807. The respective figures for input coefficients are estimated at Re. 0.10, Re. 0.001, Re. 0.005, Re. 0.26 and Re. 0.04. The agriculture crops output worth Rs. 17,274 require livestock input worth Rs. 4631, seed input valuing Rs. 676, farm land forestry inputs worth Rs. 77, CPRs-based inputs worth Rs.87, labour input Rs. 6283 and market-oriented inputs worth Rs. 501. The input coefficient per rupee of agriculture output, with regard to livestock is Re. 0.27, seed input Re. 0.04, farm forestry input Re. 0.004, CPRs input Re. 0.005, labour Re. 0.36 and purchased input Re. 0.03. Farm forestry contributed to itself Rs. 40. The total cost in case of livestock, horticulture, agriculture, farm forestry and CPRs is Rs. 22,079, 16,229, 12,255, 1450 and 7650 respectively, whereas annual production is Rs. 41,607, 44,687, 17,274, 3127 and 14,692 respectively.

Inter-dependence of Different Components of Farming Systems in Kalpa Block

Systems Linkages on Marginal Farms

Quantitative relationship between various sub-systems of farming systems in Kalpa block of Kinnaur district has been worked out category-wise and presented in this section. Abstract of flow of inputs between various activities of the system on marginal farms of Kalpa block is presented in Table 8.5. It is clear from the table that annual livestock output worth Rs. 24,527 requires crop by-products and concentrates worth Rs. 2586; farm forestry input worth Rs. 1503 and CPRs input worth Rs. 2320. Labour input worth Rs. 16,780 and purchased input amounting to Rs. 1385 is used in livestock production in this farm size category.

In terms of input-output coefficients, each rupee of current output of livestock requires agriculture input of Re. 0.11; farm land forestry input of Re. 0.06, CPRs input worth Re. 0.10. The input coefficients of labour and market-oriented inputs comes to Re. 0.68 and Re. 0.06 respectively in each rupee of livestock production.

In case of fruit crops, production of Rs. 31,329 requires an input of livestock worth Rs. 1901, farm forestry input worth Rs. 15 and CPRs input worth Rs. 174. Labour and purchased inputs work out to Rs. 5814 and Rs. 1016 respectively. The input coefficients per rupee of output of these crops are estimated at Re. 0.06, Re. 0.001, Re. 0.006, Re. 0.19 and Re. 0.03 in respect of livestock, farm forestry, CPRs, labour and purchased inputs respectively.

The annual production of agriculture crops worth Rs. 9307 utilises livestock input of Rs. 3999, seed input of its own worth Rs. 361, farm forestry worth Rs. 32 and CPRs worth Rs. 75. The value of labour used in production of agriculture is Rs. 1402 while value of market-oriented inputs is Rs. 213 only. The respective input-output coefficients work out to Re. 0.43, Re. 0.04, Re. 0.003, Re. 0.008, Re. 0.15 and Re. 0.02 in these activities of marginal farm category.

Farm forestry and community property resources make use of labour for doing harvesting operations to the extent of Rs. 1227 and Rs. 6720 respectively. The market-oriented input use is observed in farm forestry while such inputs do not exist in CPRs.

Systems Linkages on Small Farms

Transaction matrix indicating inter-dependence per household between various components, in respect of small farms of Kalpa block is summarised in Table 8.6. Like other farm sizes on small farms also, agriculture, farm land forestry and CPRs have very close relationship with livestock enterprise. The components individually provide inputs worth Rs. 3934, Rs. 1282 and Rs. 1698 for the production of total annual livestock output of Rs. 23,054. The annual labour input in livestock output turns out to be Rs. 20,040 while the value of market-oriented inputs is estimated at Rs. 1768.

In fruit crops, small farmers spend livestock input worth Rs. 3811, farm forestry input worth Rs. 48 and CPRs input worth Rs. 193. The contribution of labour inputs in production is estimated to Rs. 12,848 and purchased input value is Rs. 1333. With regard to agriculture, the livestock input contribute Rs. 5147; seed input Rs. 750; farm forestry Rs. 100; CPRs Rs. 175; labour input Rs. 7221 and market-oriented input Rs. 448. Like other farm categories share of inputs produced and used within the system is significantly high in this category also.

The input-output analysis shows that, each rupee of current livestock output use agriculture worth Re. 0.17; farm forestry worth Re. 0.06; CPRs worth Re. 0.07; labour worth Re. 0.87 and purchased inputs worth Re. 0.08. While producing one rupee from fruit crops small farmers use livestock input valued at Re. 0.07; farm forestry worth Re. 0.001; CPRs worth Re. 0.004; labour worth Re. 0.24 and market-oriented inputs worth Re. 0.03. In case of cereal crops, input-output coefficients in respect of livestock, seed input, farm forestry, CPRs, labour and purchased inputs work out to Re. 0.27, Re. 0.04, Re. 0.005, Re. 0.009, Re. 0.37 and Re. 0.02 respectively. Labour input coefficient in farm forestry and CPRs is estimated to be Re. 0.38 and Re. 0.32 respectively.

Systems Linkages on Medium Farms

The abstract of inter-sectoral flows of inputs in the farming system of medium farm category is presented in Table 8.7. A cursory glance on the table shows that in this category the yearly livestock output worth Rs. 30,408 require input of Rs. 5035 from agriculture; Rs. 696 from farm forestry; Rs. 2210 from CPRs; Rs. 20,160 from labour and Rs. 1938 worth of market-oriented inputs. The corresponding figures of input-output coefficients are estimated to be Re. 0.17, Re. 0.02, Re. 0.07, Re. 0.66 and Re. 0.06 per rupee of livestock output.

The medium farmers in Kalpa block use livestock input worth Rs. 2734; farm forestry inputs worth Rs. 21; CPRs input worth Rs. 269, labour input worth Rs. 24,010 and purchased inputs worth Rs. 2610 in the total output of Rs. 82,497 of fruit crops. In terms of input-output coefficients, livestock input is of worth Re. 0.03, farm forestry worth Re. 0.0003, CPRs is worth Re. 0.003, labour input worth Re. 0.29 and purchased inputs worth Re. 0.03.

In agriculture crops, the total annual livestock inputs in the form of farm yard manure and animal power is estimated at Rs. 5868 whose input-output coefficient works out to Re. 0.16 per rupee of output. Agriculture seed value works out to Rs. 1492 with an input coefficient of Re. 0.08 per rupee of output. Farm land forestry contributes input worth Rs. 50 whereas CPRs contribution is Rs. 182. The input-output coefficients are Re. 0.001 and Re. 0.005 respectively. Labour input value in this sector is estimated at Rs. 19,083 while value of purchased inputs is Rs. 590. The value of total annual fuelwood output works out to Rs. 1891 and Rs. 3620 from farm forestry and CPRs respectively in medium farmers. The total cost with regard to livestock, fruit crops, agriculture crops, farm forestry and CPRs works out to Rs. 30,039, Rs. 29,644, Rs. 27,265, Rs. 1540 and Rs. 6360 respectively.

Systems Linkages in Overall Situation

The dependence pattern of various components with each other, at all farm situations in Kalpa block is summarised in Table 8.8. On an average, farming system in Kalpa block has annual livestock production worth Rs. 24,932 and is utilising inputs of agriculture worth Rs. 3629; farm forestry input worth Rs. 1259; CPRs-based inputs worth Rs. 2023, labour input worth Rs. 18,842 and purchased inputs worth Rs. 1657. The input-output analysis shows that each rupee of livestock output is using agriculture input worth Re. 0.15; farm forestry input worth Re. 0.05; CPRs input worth Re. 0.08; labour input worth Re. 0.48; and purchased input worth Re. 0.07.

The fruit crops require livestock input of Rs. 2903; farm forestry input of Rs. 31, CPRs input of Rs. 200, labour input of Rs. 12,238 and purchased input of Rs. 1445. The respective figures of input-output coefficients are estimated at Re. 0.06, Re. 0.001, Re. 0.004, Re. 0.24 and Re. 0.03.

In case of agriculture crops production, the annual output worth Rs. 18921 requires livestock input of Rs. 4849; seed input of Rs. 739; farm forestry input of Rs. 66; CPRs input of Rs. 139; labour input of Rs. 7115 and purchased input of Rs. 461. The input coefficients per rupee of agriculture output, with regard to livestock comes to Re. 0.26; seed Re. 0.04; farm land forestry Re. 0.003; CPRs Re. 0.007; labour Re. 0.38 and purchased inputs worth Re. 0.02.

Farm land forestry utilises inputs worth Rs. 52 from its own production. Large amount of labour input is used by farm forestry and CPRs for the cutting of grasses, fuelwood and timber etc. The estimates of labour use in farm forestry and CPRs work out to Rs. 1284 and 5826 respectively. The labour input coefficients in farm forestry and CPRs work out to Re. 0.38 and Re. 0.33 respectively.

The total annual cost with regard to livestock, fruit crops, agriculture, farm forestry, and CPRs is estimated at Rs. 27,410, Rs. 16,817, Rs. 13,369, Rs. 1371 and Rs. 5826 respectively. The proportion of inputs produced and used within the system is quite significant. The findings that emerged from the above analysis are that on account of heavy dependence of farming systems on CPRs efforts are needed for the protection, regeneration and development of community property resources in tribal areas of Himachal Pradesh.

Inter-dependence of Different Components of Farming Systems in Nichar Block

SYSTEMS LINKAGES ON MARGINAL FARMS

Linkages between various components of farming system in Nichar block is outlined in Table 8.9. It is clear from the table that horticulture and agriculture require livestock inputs worth Rs. 1692, and Rs. 3216 respectively. Per annum livestock output worth Rs. 32,908 utilises field crops inputs (like straw, concentrates and oil cakes) worth Rs. 3049 from own farm and worth Rs. 1645 from the markets. This indicates the symbiotic relationship between crop husbandry and livestock husbandry. Farm land forestry provides grass and grazing material to the extent of Rs. 1761 to livestock component. The contribution of community property resources in livestock output when valuated at local prices work out to Rs. 3990.

In terms of input-output coefficient, each rupee of current output of livestock requires agriculture input of Re. 0.093; farm land forestry input worth Re. 0.054; common property resources worth Re. 0.121. In case of agriculture each rupee of output contains livestock input of worth Re. 0.391, seed input of Re. 0.04, farm produced other input worth Re. 0.01; and government land-based timber worth Re. 0.02. In case of horticultural crops, contribution of livestock, farm forestry and CPRs in output is Re. 0.066, Re. 0.002 and Re. 0.012 respectively per unit of output.

The dependence on farm forestry and CPRs has been observed in livestock, horticultural crops and agriculture crops. In case of farm forestry and CPRs contribution is not observed in this category. On an average, farmers make use of CPRs timber worth Rs. 40 in cowshed repair, Rs. 138 in horticultural crops and Rs. 20 in agriculture crops. Fuelwood produced on farm and collected from CPRs is directly consumed by households for domestic cooking and heating purposes. Timber from farm land forestry and CPRs is also used for house repairs by the farmers. Labour is also one of the home produced resource. The imputed value of labour works out to Rs. 23,300, Rs. 6336, Rs. 966, Rs. 1231 and Rs. 11,700 in livestock, horticultural crops, agricultural crops, farm land forestry and common property resources respectively. The purchased inputs like fertilizers, pesticides, cattle feed, transportation and threshing charges etc. have been merged in one group i.e. market-oriented inputs. The value of market oriented-inputs used is Rs. 1645 in livestock, Rs. 296 in fruit crops, Rs. 233 in agriculture and Rs. 47 in farm land forestry.

The annual value of total inputs in respect of livestock, fruit crops, agriculture, farm land forestry and CPRs individually accounts for Rs. 33,745, Rs. 8673, Rs. 4909, Rs. 1302 and Rs. 11,700 respectively.

It can be concluded from above analysis that each component of the system is heavily dependent on the input purchased on the farm itself.

Systems Linkages on Small Farms

Transaction matrix of dependence per household in respect of small farms of Nichar block is given in Table 8.10. Like marginal farms, in small farms also, agriculture, farm forestry and CPRs have very close relationship with livestock enterprise. The components individually provide inputs worth Rs. 3510, Rs. 1375 and Rs. 3275 in total annual livestock output of Rs. 32,877 in this farm category. The yearly labour input value in livestock output turns out to Rs. 20,800 while value of market-oriented inputs is estimated at Rs. 1578. In fruit crops, small farmers use livestock input worth Rs. 2983; farm forestry inputs worth Rs. 24; timber & others from CPRs worth Rs. 180. In agriculture, livestock input contribution is Rs. 4887; seed input worth Rs. 673, farm forestry worth Rs. 20, timber & others from CPRs worth Rs. 70. Like marginal farm category, share of inputs produced and used within the system in case of small farms income of this category is significantly high.

In terms of input-output coefficients, each rupee of current output of livestock requires agriculture input of Re. 0.11, farm forestry Re. 0.04, CPRs Re. 0.10. In case of fruit crops, each rupee of output contains livestock input worth Re. 0.08, farm produced and CPRs timber worth Re. 0.001 each. In agriculture production, input value of livestock, cereals itself, farm land forestry and CPRs comes to Re. 0.282, Re. 0.04, Re. 0.001 and Re. 0.004 respectively per unit of output. The value of fuelwood extracted from CPRs is Re. 2484 while fuelwood worth Re. 1313 is obtained from farm land forestry.

Systems Linkages on Medium Farms

The analysis of inter-sectoral flows of inputs on medium farm category is given in Table 8.11. A cursory glance on the table shows that in this category the annual livestock output worth Rs. 33,521 requires an input of Rs. 5202 from agriculture; Rs. 1181 from farm forestry; and Rs. 2320 from CPRs. The corresponding figures of input-output coefficient is estimated to be Re. 0.16, Re. 0.04 and Re. 0.07 per rupee of livestock output. The expenditure on labour input is Rs. 22,960 and market-oriented input is worth Rs.1896 in the production of livestock output. The contribution of livestock input, farm forestry input and CPRs input in fruit crops works out to Rs. 3185, Rs. 120 and Rs. 620 respectively. In each rupee worth of fruit crops output, livestock input is worth Re. 0.05; farm forestry worth Re. 0.002; CPRs timber & others worth Re. 0.01. The annual input value of labour and purchased inputs works out to Rs. 18,486 and Rs. 2209 respectively. In agriculture, livestock input in the form of farm yard manure and animal power is estimated to be Rs. 5447 and agriculture crops seed value at Rs. 1402. Farm land forestry contributes input worth Rs. 60, whereas CPRs contribution is Rs. 319. In terms of input-output coefficients, each rupee worth of output, requires livestock input of Re. 0.15; farm timber input of Re. 0.002, CPRs extracted timber input value of Re. 0.009. Labour input value in this sector is estimated at Rs. 15,449 and value of purchased input is Rs. 545. Labour input in farm forestry and community property resources is used for collection of fuel-wood, fodder, fruit, fibers and grazing and its value comes to be high in these sectors individually accounting for Rs. 1196 and Rs. 6540 respectively. The corresponding values of market-oriented inputs is Rs. 144 in farm land forestry and in case of CPRs there is no market purchased inputs observed in this farms size. The total cost in livestock, horticulture, agriculture, farm forestry and CPRs is Rs. 33,559, 24,620, 23,222, Rs. 1400 and Rs. 6540 respectively. It is observed that in this category, livestock system is showing a net negative balance.

SYSTEMS LINKAGES IN OVERALL SITUATION

The dependency pattern of various components with each other, at all farm situations in Nichar block under study is presented in Table 8.12. On an average, input value of cereals, farm forestry, community property resources, labour, market-oriented inputs, required for total annual livestock production worth Rs. 32,537, works out to Rs. 3618, Rs. 1466, Rs. 3355, Rs. 21,949 and Rs. 1596 respectively. In terms of input-output analysis, each rupee of livestock output is using agriculture input worth Rs. 0.11, farm forestry input worth Re. 0.045, CPRs extracted input worth Rs. 0.10. The fruit crops require livestock input worth Rs. 2612, farm land forestry input worth Rs. 44; CPRs input worth Rs. 285. The respective figures in input-output coefficients are estimated at Re. 0.07, Re. 0.001, and Re. 0.007. Labour input value works out to Rs. 10,315; purchased inputs contribution being Rs. 1074. The agriculture crops output worth Rs. 17,197 require livestock input worth Rs. 4451; seed input worth Rs. 670, farm forestry input worth Rs. 38; CPRs input worth Rs. 126. The input-output coefficients are estimated at Re. 0.26 from livestock, Re. 0.04 from its own seed, Re. 0.002 from farm forestry and Re. 0.007 from CPRs. The value of labour input in agriculture crops is Rs. 4905 and value of purchased inputs is Rs. 419. The value of input-output coefficient in production of one rupee of output is Re. 0.29; Re. 0.02 respectively. The value of fuelwood obtained from farm forestry and CPRs is Rs. 1298 and Rs. 2913 respectively. The value of timber from farm forestry and CPRs is Rs. 450 and Rs. 4644 respectively. The contribution of CPRs in the form of open grazing is valued at Rs. 3266 on the comparative ground with the market price.

Inter-dependence of Different Components of Farming Systems in Kinnaur District as a Whole

Quantitative relationship between various sub-systems of farming system in Kinnaur has been worked out category-wise in this section. The results of the analysis are discussed below:

Systems Linkages on Marginal Farms

Transaction matrix of dependence between various activities of marginal farms in Kinnaur is presented in Table 8.13. Perusal of the table shows that per annum livestock output worth Rs. 34,219 requires the input of agriculture crops worth Rs. 2561; farm land forestry worth Rs. 1418; CPRs worth Rs. 2970; labour input worth Rs. 18,005 and market purchased inputs worth Rs. 1426.

In terms of input-output coefficients, each rupee of current output of livestock requires agriculture inputs worth Re. 0.08, farm land forestry inputs worth Re. 0.04; CPRs-based inputs worth Re. 0.09; labour inputs is Re. 0.53 and market purchased inputs worth Re. 0.04. In case of fruit crops, livestock contributed inputs worth Rs. 2744; farm forestry Rs. 33; CPRs Rs. 216, labour Rs. 5443 and purchased inputs worth Rs. 1062 in annual production of Rs. 28,752 of all fruit crops. The input coefficients are Re. 0.10, Re. 0.001, Re. 0.008, Re. 0.19 and Re. 0.04 for livestock, farm forestry, CPRs, labour and purchased inputs respectively for each rupee of fruit crops output on marginal farms in Kinnaur district.

The agriculture production worth Rs. 8913 annually requires livestock-based inputs of Rs. 3851, own seed value of Rs. 345, farm forestry inputs worth Rs. 39, CPRs-based inputs Rs. 83, value of labour utilised Rs. 1303 and purchased inputs worth Rs. 231. The input-output coefficients in each rupee of agriculture production utilises livestock input of Re. 0.43, own seed Re. 0.04, farm forestry Re. 0.004, CPRs Re. 0.01, labour Re. 0.15 and purchased inputs worth Re. 0.03. The labour utilised in farm forestry and CPRs for harvesting of grass, timber and fuelwood is estimated at Rs. 1288 and Rs. 8694 respectively. The dependence of sub-systems on market-oriented inputs is relatively higher in livestock followed by fruit crops, and agriculture. The fruit being commercial crop and hence market-oriented inputs are more in these crops as compared to other activities followed by marginal farmers.

Systems Linkages on Small Farms

Transaction matrix of dependence per household in respect of small farms in Kinnaur is presented in Table 8.14. Like marginal farm size, in small farm agriculture, farm forestry and CPRs have very close relationship with livestock enterprise. The components individually provide inputs worth Rs. 3316, Rs. 1335 and Rs. 2664 in the total annual livestock output of Rs. 30,319 in this farm category. The yearly labour input value in livestock output turns out to Rs. 17,912 while value of market-oriented inputs is estimated at Rs. 1577.

In fruit crops, small farmers in Kinnaur district spend livestock input worth Rs. 3484, farm forestry input worth Rs. 32, CPRs-based inputs worth Rs. 201, labour input value is Rs. 11,428 and purchased inputs worth Rs. 1444. With regard to agriculture crops, the livestock inputs contributed Rs. 4866, seed input Rs. 726, farm forestry input Rs. 78, CPRs inputs Rs. 116, labour input worth Rs. 6452 and purchased inputs worth Rs. 460. Like marginal farm category, share of inputs produced and used within the system is significantly high in case of small farms also.

In small farm category, each rupee of current livestock output uses agriculture inputs worth Re. 0.26, farm forestry of Re. 0.004,

CPRs Re. 0.006, labour Re. 0.35 and purchased input worth Re. 0.03 respectively. While producing one rupee output of fruit crops, value of co-efficient of livestock inputs is Re. 0.07, farm forestry Re. 0.001, CPRs, Re. 0.004, labour Re. 0.24 and purchased inputs Re. 0.03. In case of agriculture, input-output coefficients in respect of livestock, agriculture, farm forestry, CPRs, labour and purchased input works to Re. 0.261, Re. 0.04, Re. 0.004, Re. 0.006, Re. 0.35 and Re. 0.03 respectively.

Systems Linkages in Medium Farms

Per farm quantitative relationship between various activities of farming system on medium farms of Kinnaur district is summarised in Table 8.15. It may be seen from this table that livestock output worth Rs. 35,625 per annum requires an input of agriculture crops worth Rs. 5080 per annum. Value of this input per rupee of livestock output works out to Re. 0.14. The contribution of fodder and grasses etc. from farm forestry and CPRs comes to Rs. 829 and Rs. 2306 respectively per annum which, in terms of input coefficient per rupee of livestock output is estimated at Re. 0.03 and Re. 0.07 respectively. Fruit crops output worth Rs. 76,200 requires inputs of livestock worth Rs. 3747, farm forestry Rs. 52, CPRs Rs. 424, labour input worth Rs. 22,231 and purchased inputs Rs. 2686. The input coefficient per rupee of output of these crops works out to Re. 0.05 of livestock inputs, Re. 0.001 of farm forestry, Re. 0.006 of CPRs, Re. 0.30 of labour and Re. 0.04 of purchased inputs.

The annual production of agriculture crops worth Rs. 33,713 utilises livestock input of Rs. 5933, own seed input worth Rs. 1486; farm forestry input of Rs. 52; CPRs input worth Rs. 214; labour input of Rs. 17,711 and market-oriented inputs worth Rs. 601. The respective input-output coefficient works out to Re. 0.16, Re. 0.04, Re.0.001, Re. 0.006, Re. 0.47 and Re. 0.02 in medium farm group.

Vegetable farming being a weak activity in Kinnaur farming system, utilises negligible amount of inputs from other sectors in the area under study. Farm land forestry and CPRs make use of labour for doing harvesting operations to the extent of Rs. 1344 and Rs. 6917 respectively.

Systems Linkages in Overall Situation

The dependence pattern of various components with each other, at all farm situation in Kinnaur district is given in Table 8.16. On an average, in the tribal farming system, annual livestock production worth Rs. 32,473 is utilising inputs of agriculture worth Rs. 3239, farm land forestry inputs worth Rs. 1335, CPRs-based inputs Rs. 2762; labour inputs worth Rs. 18,404 and purchased inputs worth Rs. 1563. The input-output analysis shows that each rupee of livestock output is using agriculture inputs worth Re. 0.10; farm forestry inputs worth Re. 0.04; CPRs inputs Re. 0.09; labour input worth Re. 0.57 and purchased inputs worth Re. 0.05.

The fruit crops require livestock input of Rs. 3255; farm forestry inputs of Rs. 36; CPRs produced inputs worth Rs. 238; labour input of Rs. 10,860 and purchased inputs of Rs. 1432. The respective figures of input-output coefficients are estimated at Re. 0.07, Re. 0.001, Re. 0.005, Re. 0.24 and Re. 0.03.

In case of agriculture crops production, the annual output worth Rs. 17,856 requires livestock input of Rs. 4652; seed input of Rs. 697; farm forestry of Rs. 60; CPRs of Rs. 118; labour of Rs. 6139; purchased inputs of Rs. 459. The input coefficients per rupee of agriculture production with regard to livestock comes to Re. 0.26, seed Re. 0.04; farm forestry Re. 0.003, CPRs inputs worth Re. 0.007, labour Re. 0.34 and purchased inputs worth Re. 0.03.

Farm land forestry utilises worth Rs. 42 inputs from its own production. The estimates of labour use in farm forestry sector and CPRs work out to Rs. 1327 and Rs. 7683 respectively. The labour input coefficient in farm forestry works out to Re. 0.40 while in CPRs it is Re. 0.48.

The total annual cost with regard to livestock, fruit crops, agriculture crop, farm forestry and CPRs is estimated at Rs. 27,303, Rs. 15,821, Rs. 12,125, Rs. 1409 and Rs. 7683 respectively. The proportion of inputs produced and used within the system is again quite significant. The inter-sectoral linkage is significantly high between agriculture and livestock and it is observed to be weak in case of vegetable farming and horticultural economy in the study area.

Table 8.1. Transaction matrix dependence per household between various components of farming systems on marginal farms of Pooh block

(Rs./Farm/Annum)

Producing sectors	Consuming sectors					
	Live-stock	Horti-culture	Agri-culture	Farm forestry	CPRs	Total produc-tion
1. Livestock	**–**	**4609 (.162)**	**4221 (.466)**	**–**	**–**	**46464**
a) Sheep & goats	–	1000	340	–	–	25550
b) Others	–	3609	3881	–	–	20914
2. Horticulture	**–**	**–**	**–**	**–**	**–**	**28515**
a) Apple	–	–	–	–	–	24730
b) Others	–	–	–	–	–	3785
3. Agriculture	**2117 (.046)**	**–**	**352 (.039)**	**–**	**–**	**9053**
a) Grass	717	–	–	–	–	717
b) Concentrate	1400	–	–	–	–	1400
c) Food crops	–	–	–	–	–	6936
4. Farm forestry	**1028 (.022)**	**47.0 (.002)**	**30.0 (.003)**	**22.0 (.009)**	**–**	**2394**
a) Timber wood	–	–	–	–	–	264
b) Fuelwood	–	–	–	–	–	810
c) Grass & grazing	1008	–	–	–	–	1008
d) Others	20.0	47.0	30.0	22.0	–	312
5. CPRs	**2847 (.061)**	**186 (.007)**	**60.0 (.007)**	**–**	**–**	**15176**
a) Chilgoza	–	–	–	–	–	815
b) Timber wood	40.0	164	40.0	–	–	2416
c) Fuelwood	–	–	–	–	–	2067
d) Grass & grazing	2800	–	–	–	–	9600
e) Others	7.0	22.0	20.0	–	–	278
6. Labour (domes-tic & hired)	**14900 (.321)**	**4678 (.164)**	**1477 (.163)**	**1408 (.588)**	**8400 (.554)**	**30863**
7. Market Pur-chased inputs	**1288 (.028)**	**1343 (.047)**	**250 (.028)**	**47 (.020)**	**–**	**2928**
Total costs	**22180**	**10863**	**6390**	**1477**	**8400**	**49310**
Net return	**24284**	**17652**	**2663**	**917**	**6776**	**52292**

Note: Figures in parentheses denote the input-output co-efficient.

Table 8.2. Transaction matrix dependence per household between various components of farming systems on small farms of Pooh block

(Rs./Farm/Annum)

Producing sectors	Consuming sectors					
	Live-stock	Horti-culture	Agri-culture	Farm forestry	CPRs	Total produc-tion
1. Livestock	–	**3710 (.072)**	**4544 (.233)**	–	–	**35046**
a) Sheep & goats	–	500	481	–	–	19508
b) Others	–	3210	4063	–	–	15538
2. Horticulture	–	–	–	–	–	**51742**
a) Apple	–	–	–	–	–	44560
b) Others	–	–	–	–	–	7182
3. Agriculture	**2440 (.070)**	–	**762 (.039)**	–	–	**19502**
a) Grass	1472	–	–	–	–	1472
b) Concentrate	968	–	–	–	–	968
c) Food Crops	–	–	–	–	–	17062
4. Farm Forestry	**1346 (.038)**	**26 (.001)**	**120 (.006)**	**40 (.011)**	–	**3650**
a) Timber wood	–	–	–	–	–	576
b) Fuelwood	–	–	–	–	–	1582
c) Grass & grazing	1316	–	–	–	–	1316
d) Others	30	26	120	40	–	176
5. CPRs	**2988 (.085)**	**234 (.005)**	**106 (.005)**	–	–	**13977**
a) Chilgoza	–	–	–	–	–	1089
b) Timber wood	69	29	21	–	–	2850
c) Fuelwood	–	–	–	–	–	2056
d) Grass & grazing	2880	–	–	–	–	7330
e) Others	39	205	85	–	–	652
6. Labour (domestic & hired)	**12359 (.353)**	**11149 (.215)**	**8059 (.413)**	**1292 (.354)**	**6840 (.489)**	**39689**
7. Market Purchased inputs	**1375 (.039)**	**1831 (.035)**	**503 (.026)**	**26 (.007)**	–	**3735**
Total costs	**20508**	**16950**	**14094**	**1358**	**6840**	**59750**
Net return	**14538**	**34792**	**5408**	**2292**	**7137**	**64167**

Note: Figures in parentheses denote the input-output co-efficient.

Table 8.3. Transaction matrix dependence per household between various components of farming systems on medium farms of Pooh block

(Rs./Farm/Annum)

Producing sectors	Consuming sectors					
	Live-stock	Horti-culture	Agri-culture	Farm forestry	CPRs	Total produc-tion
1. Livestock	–	**7309 (.096)**	**6985 (.169)**	–	–	**58030**
a) Sheep & goats	–	1200	650	–	–	33871
b) Others	–	6109	6335	–	–	24159
2. Horticulture	–	–	–	–	–	**76309**
a) Apple	–	–	–	–	–	60761
b) Others	–	–	–	–	–	15548
3. Agriculture	**4976 (.086)**	–	**1627 (.039)**	–	–	**41275**
a) Grass	3712	–	–	–	–	3712
b) Concentrate	1264	–	–	–	–	1264
c) Food crops	–	–	–	–	–	36299
4. Farm forestry	**740 (.013)**	**7 (.0001)**	**40 (.001)**	**123 (.036)**	–	**3377**
a) Timber wood	–	–	–	–	–	792
b) Fuelwood	–	–	–	–	–	1840
c) Grass & grazing	700	–	–	–	–	700
d) Others	40	7	40	123	–	45
5. CPRs	**3160 (.054)**	**452 (.006)**	**99 (.002)**	–	–	**16684**
a) Chilgoza	–	–	–	–	–	2917
b) Timber wood	50	152	20	–	–	1162
c) Fuelwood	–	–	–	–	–	2340
d) Grass & grazing	3000	–	–	–	–	9500
e) Others	110	300	79	–	–	765
6. Labour (domes-tic & hired)	**19801 (.341)**	**24649 (.323)**	**18448 (.447)**	**1658 (.491)**	**9000 (.539)**	**73556**
7. Market Pur-chased inputs	**2115 (.036)**	**3750 (.049)**	**717 (.017)**	**81 (.024)**	–	**6663**
Total costs	**30792**	**36167**	**27916**	**1862**	**9000**	**105737**
Net return	**27238**	**40142**	**13359**	**1515**	**7684**	**89938**

Note: Figures in parentheses denote the input-output co-efficient.

Table 8.4. Transaction matrix dependence per household between various components of farming systems on farms of Pooh block

(Rs./Farm/Annum)

Producing sectors	Consuming sectors					
	Live-stock	Horti-culture	Agri-culture	Farm forestry	CPRs	Total produc-tion
1. Livestock	–	**4384 (.098)**	**4631 (.268)**	–	–	**41607**
a) Sheep & goats	–	900	300	–	–	23174
b) Others	–	3484	4331	–	–	18433
2. Horticulture	–	–	–	–	–	**44687**
a) Apple	–	–	–	–	–	38116
b) Others	–	–	–	–	–	6571
3. Agriculture	**2536 (.061)**	–	**676 (.039)**	–	–	**17274**
a) Grass	1370	–	–	–	–	1370
b) Concentrate	1166	–	–	–	–	1166
c) Food crops	–	–	–	–	–	14738
4. Farm forestry	**1165 (.028)**	**33 (.001)**	**77 (.004)**	**40 (.013)**	–	**3127**
a) Timber wood	–	–	–	–	–	471
b) Fuelwood	–	–	–	–	–	1298
c) Grass & grazing	1139	–	–	–	–	1139
d) Others	26	33	77	40	–	218
5. CPRs	**2948 (.071)**	**234 (.005)**	**87 (.005)**	–	–	**14692**
a) Chilgoza	–	–	–	–	–	1142
b) Timber wood	61	83	75	–	–	2529
c) Fuelwood	–	–	–	–	–	2086
d) Grass & grazing	2859	–	–	–	–	8423
e) Others	28	151	12	–	–	512
6. Labour (domestic & hired)	**14016 (.337)**	**9771 (.259)**	**6283 (.364)**	**1382 (.442)**	**7650 (.521)**	**30308**
7. Market purchased inputs	**1414 (.03)**	**1807 (.040)**	**501 (.029)**	**28 (.009)**	–	**3750**
Total costs	**22079**	**16229**	**12255**	**1450**	**7650**	**59663**
Net return	**19528**	**28458**	**5019**	**1677**	**7042**	**61724**

Note: Figures in parentheses denote the input-output co-efficient.

Table 8.5. Transaction matrix dependence per household between various components of farming systems on marginal farms of Kalpa block

(Rs./Farm/Annum)

Producing sectors	Consuming sectors					
	Livestock	Horticulture	Agriculture	Farm forestry	CPRs	Total production
1. Livestock	–	**1901 (.061)**	**3999 (.430)**	–	–	**24527**
a) Sheep & goats	–	600	438	–	–	11239
b) Others	–	1301	3561	–	–	13288
2. Horticulture	–	–	–	–	–	**31329**
a) Apple	–	–	–	–	–	25850
b) Others	–	–	–	–	–	2779
3. Agriculture	**2586 (.105)**	–	**361 (.039)**	–	–	**9307**
a) Grass	786	–	–	–	–	786
b) Concentrate	1800	–	–	–	–	1800
c) Food crops	–	–	–	–	–	6721
4. Farm forestry	**1503 (.061)**	**15 (.001)**	**32 (.003)**	**18 (.005)**	–	**3286**
a) Timber wood	–	–	–	–	–	720
b) Fuelwood	–	–	–	–	–	987
c) Grass & grazing	1463	–	–	–	–	1463
d) Others	40	15	32	18	–	116
5. CPRs	**2320 (.095)**	**174 (.006)**	**75 (.008)**	–	–	**17187**
a) Chilgoza	–	–	–	–	–	684
b) Timber wood	30	136	35	–	–	5566
c) Fuelwood	–	–	–	–	–	3399
d) Grass & grazing	2240	–	–	–	–	7190
e) Others	50	38	40	–	–	348
6. Labour (domestic & hired)	**16780 (.684)**	**5814 (.186)**	**1402 (.151)**	**1227 (.373)**	**6720 (.391)**	**31943**
7. Market purchased inputs	**1385 (.056)**	**1016 (.032)**	**213 (.023)**	**36 (.011)**	–	**2650**
Total costs	**24574**	**8920**	**6082**	**1281**	**6720**	**47577**
Net return	**-47**	**22409**	**3225**	**2005**	**10467**	**38059**

Note: Figures in parentheses denote the input-output co-efficient.

Table 8.6. Transaction matrix dependence per household between various components of farming systems on small farms of Kalpa block

(Rs./Farm/Annum)

Producing sectors	Consuming sectors					
	Live-stock	Horti-culture	Agri-culture	Farm forestry	CPRs	Total produc-tion
1. Livestock	–	**3811 (.072)**	**5147 (.266)**	–	–	**23054**
a) Sheep & goats	–	800	238	–	–	8423
b) Others	–	3011	4909	–	–	14631
2. Horticulture	–	–	–	–	–	**53141**
a) Apple	–	–	–	–	–	51030
b) Others	–	–	–	–	–	2111
3. Agriculture	**3934 (.171)**	–	**750 (.039)**	–	–	**19342**
a) Grass	1686	–	–	–	–	1686
b) Concentrate	2248	–	–	–	–	2248
c) Food crops	–	–	–	–	–	15408
4. Farm forestry	**1282 (.056)**	**48 (.001)**	**100 (.005)**	**41 (.012)**	–	**3394**
a) Timber wood	–	–	–	–	–	389
b) Fuelwood	–	–	–	–	–	1436
c) Grass & grazing	1246	–	–	–	–	1246
d) Others	36	48	100	41	–	232
5. CPRs	**1698 (.074)**	**193 (.004)**	**175 (.009)**	–	–	**15149**
a) Chilgoza	–	–	–	–	–	844
b) Timber wood	48	104	55	–	–	5374
c) Fuelwood	–	–	–	–	–	2704
d) Grass & grazing	1620	–	–	–	–	5470
e) Others	30	89	120	–	–	757
6. Labour (domestic & hired)	**20040 (.869)**	**12848 (.242)**	**7221 (.373)**	**1276 (.376)**	**4860 (.321)**	**46245**
7. Market purchased inputs	**1768 (.077)**	**1333 (.025)**	**448 (.023)**	**62 (.018)**	–	**3611**
Total costs	**28722**	**18233**	**13841**	**1379**	**4860**	**67035**
Net return	**-5668**	**34908**	**5501**	**2015**	**10289**	**47045**

Note: Figures in parentheses denote the input-output co-efficient.

Table 8.7. Transaction matrix dependence per household between various components of farming systems on medium farms of Kalpa block

(Rs./Farm/Annum)

Producing sectors	Consuming sectors					
	Live-stock	Horti-culture	Agri-culture	Farm forestry	CPRs	Total produc-tion
1. Livestock	–	**2734 (.033)**	**5868 (.155)**	–	–	**30408**
a) Sheep & goats	–	600	295	–	–	12053
b) Others	–	2134	5573	–	–	18355
2. Horticulture	–	–	–	–	–	**82497**
a) Apple	–	–	–	–	–	76000
b) Others	–	–	–	–	–	6497
3. Agriculture	**5035 (.166)**	–	**1492 (.084)**	–	–	**37754**
a) Grass	3163	–	–	–	–	3163
b) Concentrate	1872	–	–	–	–	1872
c) Food crops	–	–	–	–	–	32719
4. Farm forestry	**696 (.023)**	**21 (.0003)**	**50 (.001)**	**152 (.045)**	–	**3414**
a) Timber wood	–	–	–	–	–	720
b) Fuelwood	–	–	–	–	–	1891
c) Grass & grazing	651	–	–	–	–	651
d) Others	45	21	50	152	–	152
5. CPRs	**2210 (.073)**	**269 (.003)**	**182 (.005)**	–	–	**24357**
a) Chilgoza	–	–	–	–	–	1333
b) Timber wood	70	189	80	–	–	9924
c) Fuelwood	–	–	–	–	–	3620
d) Grass & grazing	2120	–	–	–	–	9190
e) Others	20	80	102	–	–	290
6. Labour (domestic & hired)	**20160 (.663)**	**24010 (.291)**	**19083 (.505)**	**1313 (.385)**	**6360 (.261)**	**70926**
7. Market purchased inputs	**1938 (.064)**	**2610 (.032)**	**590 (.016)**	**75 (.022)**	–	**5213**
Total costs	**30039**	**29644**	**27265**	**1540**	**6360**	**94848**
Net return	**369**	**52853**	**10489**	**1874**	**17997**	**83582**

Note: Figures in parentheses denote the input-output co-efficient.

Table 8.8. Transaction matrix dependence per household between various components of farming systems on farms of Kalpa block

(Rs./Farm/Annum)

Producing sectors	Consuming sectors					
	Live-stock	Horti-culture	Agri-culture	Farm forestry	CPRs	Total produc-tion
1. Livestock	–	**2903 (.058)**	**4849 (.256)**	–	–	**24932**
a) Sheep & goats	–	600	341	–	–	10130
b) Others	–	2303	4508	–	–	14802
2. Horticulture	–	–	–	–	–	**50300**
a) Apple	–	–	–	–	–	47147
b) Others	–	–	–	–	–	3153
3. Agriculture	**3629 (.146)**	–	**739 (.039)**	–	–	**18921**
a) Grass	1616	–	–	–	–	1616
b) Concentrate	2013	–	–	–	–	2013
c) Food crops	–	–	–	–	–	15292
4. Farm forestry	**1259 (.050)**	**31 (.001)**	**66 (.003)**	**52 (.015)**	–	**3357**
a) Timber wood	–	–	–	–	–	572
b) Fuelwood	–	–	–	–	–	1350
c) Grass & grazing	1220	–	–	–	–	1220
d) Others	39	31	66	52	–	215
5. CPRs	**2023 (.081)**	**200 (.004)**	**139 (.007)**	–	–	**17574**
a) Chilgoza	–	–	–	–	–	873
b) Timber wood	35	131	26	–	–	6268
c) Fuelwood	–	–	–	–	–	2129
d) Grass & grazing	1942	–	–	–	–	6772
e) Others	46	69	113	–	–	1532
6. Labour (domestic & hired)	**18842 (.475)**	**12238 (.243)**	**7115 (.376)**	**1284 (.382)**	**5826 (.332)**	**45305**
7. Market purchased inputs	**1657 (.066)**	**1445 (.029)**	**461 (.024)**	**35 (.010)**	—	**3598**
Total costs	**27410**	**16817**	**13369**	**1371**	**5826**	**64793**
Net return	**-2478**	**33483**	**5552**	**1985**	**11749**	**50291**

Note: Figures in parentheses denote the input-output co-efficient.

Table 8.9. Transaction matrix dependence per household between various components of farming systems on marginal farms of Nichar block

(Rs./Farm/Annum)

Producing sectors	Consuming sectors					
	Live-stock	Horti-culture	Agri-culture	Farm forestry	CPRs	Total produc-tion
1. Livestock	**–**	**1692 (.066)**	**3216 (.391)**	**–**	**–**	**32908**
a) Sheep & goats	–	600	290	–	–	19896
b) Others	–	1092	2926	–	–	13012
2. Horticulture	**–**	**–**	**–**	**–**	**–**	**25555**
a) Apple	–	–	–	–	–	18560
b) Others	–	–	–	–	–	6995
3. Agriculture	**3049 (.093)**	**–**	**314 (.038)**	**–**	**–**	**8217**
a) Grass	649	–	–	–	–	649
b) Concentrate	2400	–	–	–	–	2400
c) Food crops	–	–	–	–	–	5168
4. Farm forestry	**1761 (.054)**	**41 (.002)**	**60 (.007)**	**24 (.006)**	**–**	**3718**
a) Timber wood	–	–	–	–	–	156
b) Fuelwood	–	–	–	–	–	1044
c) Grass & grazing	1701	–	–	–	–	1701
d) Others	60	41	60	24	–	816
5. CPRs	**3990 (.121)**	**308 (.012)**	**120 (.015)**	**–**	**–**	**16178**
a) Chilgoza	–	–	–	–	–	–
b) Timber wood	40	138	20	–	–	3794
c) Fuelwood	–	–	–	–	–	3199
d) Grass & grazing	3900	–	–	–	–	8550
e) Others	50	170	100	–	–	635
6. Labour (domestic & hired)	**23300 (.708)**	**6336 (.248)**	**966 (.118)**	**1231 (.331)**	**11700 (.723)**	**43533**
7. Market purchased inputs	**1645 (.050)**	**296 (.012)**	**233 (.028)**	**47 (.013)**	**–**	**2221**
Total costs	**33745**	**8673**	**4909**	**1302**	**11700**	**60329**
Net return	**-837**	**16882**	**3308**	**2416**	**4478**	**26247**

Note: Figures in parentheses denote the input-output co-efficient.

Table 8.10. Transaction matrix dependence per household between various components of farming systems on small farms of Nichar block

(Rs./Farm/Annum)

Producing sectors	Consuming sectors					
	Live-stock	Horti-culture	Agri-culture	Farm forestry	CPRs	Total produc-tion
1. Livestock	–	**2983 (.076)**	**4887 (.282)**	–	–	**32877**
a) Sheep & goats	–	800	177	–	–	20146
b) Others	–	2183	4710	–	–	12731
2. Horticulture	–	–	–	–	–	**39138**
a) Apple	–	–	–	–	–	35600
b) Others	–	–	–	–	–	3538
3. Agriculture	**3510 (.107)**	–	**673 (.039)**	–	–	**17269**
a) Grass	1518	–	–	–	–	1518
b) Concentrate	1992	–	–	–	–	1992
c) Food crops	–	–	–	–	–	13759
4. Farm forestry	**1375 (.042)**	**24 (.001)**	**20 (.001)**	**32 (.004)**	–	**3403**
a) Timber wood	–	–	–	–	–	624
b) Fuelwood	–	–	–	–	–	1313
c) Grass & grazing	1330	–	–	–	–	1330
d) Others	45	24	20	32	–	136
5. CPRs	**3275 (.100)**	**180 (.005)**	**70 (.004)**	–	–	**15218**
a) Chilgoza	–	–	–	–	–	–
b) Timber wood	35	121	40	–	–	5260
c) Fuelwood	–	–	–	–	–	2484
d) Grass & grazing	3200	–	–	–	–	6700
e) Others	40	59	30	–	–	774
6. Labour (domes-tic & hired)	**20800 (.633)**	**10356 (.265)**	**4335 (.251)**	**1414 (.416)**	**9600 (.631)**	**46505**
7. Market pur-chased inputs	**1578 (.048)**	**1209 (.031)**	**434 (.025)**	**38 (.011)**	–	**3259**
Total costs	**30538**	**14752**	**10419**	**1484**	**9600**	**66793**
Net return	**2339**	**24386**	**6850**	**1919**	**5618**	**41112**

Note: Figures in parentheses denote the input-output co-efficient.

Table 8.11. Transaction matrix dependence per household between various components of farming systems on medium farms of Nichar block

(Rs./Farm/Annum)

Producing sectors	Consuming sectors					
	Livestock	Horticulture	Agriculture	Farm forestry	CPRs	Total production
1. Livestock	–	**3185 (.047)**	**5447 (.153)**	–	–	**33521**
a) Sheep & goats	–	700	350	–	–	17441
b) Others	–	2485	5097	–	–	16080
2. Horticulture	–	–	–	–	–	**67554**
a) Apple	–	–	–	–	–	61670
b) Others	–	–	–	–	–	5884
3. Agriculture	**5202 (.155)**	–	**1402 (.039)**	–	–	**35714**
a) Grass	2818	–	–	–	–	2818
b) Concentrate	2384	–	–	–	–	2384
c) Food crops	–	–	–	–	–	30512
4. Farm forestry	**1181 (.035)**	**120 (.002)**	**60 (.002)**	**40 (.011)**	–	**3562**
a) Timber wood	–	–	–	–	–	432
b) Fuelwood	–	–	–	–	–	1769
c) Grass & grazing	1141	–	–	–	–	1141
d) Others	40	120	60	40	–	222
5. CPRs	**2320 (.069)**	**620 (.009)**	**319 (.009)**	–	–	**17298**
a) Chilgoza	–	–	–	–	–	–
b) Timber wood	60	140	40	–	–	4182
c) Fuelwood	–	–	–	–	–	3886
d) Grass & grazing	2180	–	–	–	–	8430
e) Others	80	480	279	–	–	800
6. Labour (domestic & hired)	**22960 (.685)**	**18486 (.274)**	**15449 (.433)**	**1196 (.336)**	**6540 (.378)**	**64631**
7. Market Purchased inputs	**1896 (.057)**	**2209 (.033)**	**545 (.015)**	**144 (.040)**	–	**4794**
Total costs	**33559**	**24620**	**23222**	**1400**	**6540**	**89341**
Net return	**-38**	**42934**	**12492**	**2162**	**10758**	**68308**

Note: Figures in parentheses denote the input-output co-efficient.

Table 8.12. Transaction matrix dependence per household between various components of farming systems on farms of Nichar block

(Rs./Farm/Annum)

Producing sectors	Consuming sectors					
	Live-stock	Horti-culture	Agri-culture	Farm forestry	CPRs	Total produc-tion
1. Livestock	–	**2612 (.067)**	**4451 (.259)**	–	–	**32537**
a) Sheep & goats	–	500	462	–	–	19221
b) Others	–	2112	3989	–	–	13316
2. Horticulture	–	–	–	–	–	**39141**
a) Apple	–	–	–	–	–	34179
b) Others	–	–	–	–	–	4962
3. Agriculture	**3618 (.111)**	–	**670 (.039)**	–	–	**17197**
a) Grass	1441	–	–	–	–	1441
b) Concentrate	2177	–	–	–	–	2177
c) Food crops	–	–	–	–	–	13579
4. Farm forestry	**1466 (.045)**	**44 (.001)**	**38 (.002)**	**31 (.009)**	–	**3525**
a) Timber wood	–	–	–	–	–	450
b) Fuelwood	–	–	–	–	–	1298
c) Grass & grazing	1417	–	–	–	–	1417
d) Others	49	44	38	31	–	360
5. CPRs	**3355 (.103)**	**285 (.007)**	**126 (.007)**	–	–	**15826**
a) Chilgoza	–	–	–	–	–	–
b) Timber wood	60	125	40	–	–	4644
c) Fuelwood	–	–	–	–	–	2913
d) Grass & grazing	3266	–	–	–	–	7532
e) Others	29	160	86	–	–	737
6. Labour (domestic & hired)	**21949 (.675)**	**10315 (.264)**	**4905 (.285)**	**1327 (.376)**	**9798 (.619)**	**48294**
7. Market purchased inputs	**1596 (.049)**	**1074 (.027)**	**419 (.024)**	**57 (.016)**	–	**3146**
Total costs	**31984**	**14330**	**10609**	**1415**	**9798**	**68136**
Net return	**553**	**24811**	**6588**	**2110**	**6028**	**40090**

Note: Figures in parentheses denote the input-output co-efficient.

Table 8.13. Transaction matrix dependence per household between various components of farming systems on marginal farms of Kinnaur district

(Rs./Farm/Annum)

Producing sectors	Consuming sectors					
	Live-stock	Horti-culture	Agri-culture	Farm forestry	CPRs	Total produc-tion
1. Livestock	–	**2744 (.095)**	**3851 (.432)**	–	–	**34219**
a) Sheep & goats	–	733	363	–	–	18467
b) Others	–	2011	3488	–	–	15752
2. Horticulture	–	–	–	–	–	**28752**
a) Apple	–	–	–	–	–	24440
b) Others	–	–	–	–	–	4312
3. Agriculture	**2561 (.075)**	–	**345 (.039)**	–	–	**8913**
a) Grass	724	–	–	–	–	724
b) Concentrate	1837	–	–	–	–	1837
c) Food crops	–	–	–	–	–	6352
4. Farm forestry	**1418 (.041)**	**33 (.001)**	**39 (.004)**	**21 (.007)**	–	**3111**
a) Timber wood	–	–	–	–	–	408
b) Fuelwood	–	–	–	–	–	944
c) Grass & grazing	1379	–	–	–	–	1379
d) Others	39	33	39	21	–	380
5. CPRs	**2970 (.087)**	**216 (.008)**	**83 (.009)**	–	–	**16230**
a) Chilgoza	–	–	–	–	–	533
b) Timber wood	36	146	32	–	–	4013
c) Fuelwood	–	–	–	–	–	2898
d) Grass & grazing	2898	–	–	–	–	8379
e) Others	36	70	51	–	–	407
6. Labour (domestic & hired)	**18005 (.526)**	**5443 (.189)**	**1303 (.146)**	**1288 (.414)**	**8694 (.536)**	**34733**
7. Market purchased inputs	**1426 (.042)**	**1062 (.037)**	**231 (.026)**	**43 (.014)**	–	**2762**
Total costs	**26380**	**9498**	**5852**	**1352**	**8694**	**51776**
Net return	**7839**	**19254**	**3061**	**1759**	**7536**	**39449**

Note: Figures in parentheses denote the input-output co-efficient.

Table 8.14. Transaction matrix dependence per household between various components of farming systems on small farms of Kinnaur district

(Rs./Farm/Annum)

Producing sectors	Consuming sectors					
	Live-stock	Horti-culture	Agri-culture	Farm forestry	CPRs	Total produc-tion
1. Livestock	**–**	**3484 (.073)**	**4866 (.261)**	**–**	**–**	**30319**
a) Sheep & goats	–	706	292	–	–	16083
b) Others	–	2778	4574	–	–	14236
2. Horticulture	**–**	**–**	**–**	**–**	**–**	**47703**
a) Apple	–	–	–	–	–	43497
b) Others	–	–	–	–	–	4206
3. Agriculture	**3316 (.109)**	**–**	**726 (.039)**	**–**	**–**	**18652**
a) Grass	1559	–	–	–	–	1559
b) Concentrate	1757	–	–	–	–	1757
c) Food crops	–	–	–	–	–	15336
4. Farm forestry	**1335 (.044)**	**32 (.001)**	**78 (.004)**	**38 (.011)**	**–**	**3477**
a) Timber wood	–	–	–	–	–	531
b) Fuelwood	–	–	–	–	–	1438
c) grass & grazing	1298	–	–	–	–	1298
d) Others	37	32	78	38	–	210
5. CPRs	**2664 (.088)**	**201 (.004)**	**116 (.006)**	**–**	**–**	**14807**
a) Chilgoza	–	–	–	–	–	619
b) Timber wood	50	80	39	–	–	4545
c) Fuelwood	–	–	–	–	–	2422
d) Grass & grazing	2578	–	–	–	–	6491
e) Others	36	121	77	–	–	730
6. Labour (domestic & hired)	**17912 (.591)**	**11428 (.240)**	**6452 (.346)**	**1330 (.383)**	**7172 (.484)**	**44294**
7. Market purchased inputs	**1577 (.052)**	**1444 (.030)**	**460 (.025)**	**42 (.012)**	**–**	**3523**
Total costs	**26804**	**16589**	**12698**	**1410**	**7172**	**64673**
Net return	**3515**	**31114**	**5954**	**2067**	**7635**	**50285**

Note: Figures in parentheses denote the input-output co-efficient.

Table 8.15. Transaction matrix dependence per household between various components of farming systems on medium farms of Kinnaur district

(Rs./Farm/Annum)

Producing sectors	Consuming sectors					
	Live-stock	Horti-culture	Agri-culture	Farm forestry	CPRs	Total produc-tion
1. Livestock	**–**	**3747 (.049)**	**5933 (.157)**	**–**	**–**	**35625**
a) Sheep & goats	–	747	380	–	–	16964
b) Others	–	3000	5553	–	–	18661
2. Horticulture	**–**	**–**	**–**	**–**	**–**	**76200**
a) Apple	–	–	–	–	–	68217
b) Others	–	–	–	–	–	7983
3. Agriculture	**5080 (.143)**	**–**	**1486 (.039)**	**–**	**–**	**33713**
a) Grass	3146	–	–	–	–	3146
b) Concentrate	1934	–	–	–	–	1934
c) Food crops	–	–	–	–	–	32633
4. Farm forestry	**871 (.024)**	**52 (.001)**	**52 (.001)**	**108 (.031)**	**–**	**3458**
a) Timber wood	–	–	–	–	–	635
b) Fuelwood	–	–	–	–	–	1839
c) Grass & grazing	829	–	–	–	–	829
d) Others	42	52	52	108	–	155
5. CPRs	**2466 (.069)**	**424 (.006)**	**214 (.006)**	**–**	**–**	**20490**
a) Chilgoza	–	–	–	–	–	1172
b) Timber wood	63	165	55	–	–	6307
c) Fuelwood	–	–	–	–	–	3471
d) Grass & grazing	2306	–	–	–	–	8954
e) Others	97	259	159	–	–	586
6. Labour (domestic & hired)	**21016 (.590)**	**22231 (.292)**	**17711 (.470)**	**1344 (.387)**	**6917 (.338)**	**69219**
7. Market purchased inputs	**1957 (.055)**	**2686 (.035)**	**601 (.016)**	**100 (.029)**	**–**	**5344**
Total costs	**31390**	**29140**	**25997**	**1552**	**6917**	**94996**
Net return	**4235**	**47060**	**7716**	**1906**	**13573**	**78492**

Note: Figures in parentheses denote the input-output co-efficient.

Table 8.16. Transaction matrix dependence per household between various components of farming systems on farms of Kinnaur district

(Rs./Farm/Annum)

Producing sectors	Consuming sectors					
	Live-stock	Horti-culture	Agri-culture	Farm forestry	CPRs	Total produc-tion
1. Livestock	–	**3255 (.072)**	**4652 (.261)**	–	–	**32473**
a) Sheep & goats	–	722	368	–	–	17062
b) Others	–	2533	4284	–	–	15411
2. Horticulture	–	–	–	–	–	**44933**
a) Apple	–	–	–	–	–	40152
b) Others	–	–	–	–	–	4781
3. Agriculture	**3239 (.100)**	–	**697 (.039)**	–	–	**17856**
a) Grass	1428	–	–	–	–	1428
b) Concentrate	1811	–	–	–	–	1811
c) Food crops	–	–	–	–	–	14617
4. Farm forestry	**1335 (.041)**	**36 (.001)**	**60 (.003)**	**42 (.013)**	–	**3343**
a) Timber wood	–	–	–	–	–	502
b) Fuelwood	–	–	–	–	–	1317
c) Grass & grazing	1260	–	–	–	–	1260
d) Others	18	36	60	42	–	264
5. CPRs	**2762 (.085)**	**238 (.005)**	**118 (.007)**	–	–	**16128**
a) Chilgoza	–	–	–	–	–	667
b) Timber wood	51	115	45	–	–	4604
c) Fuelwood	–	–	–	–	–	2743
d) Grass & grazing	2654	–	–	–	–	7520
e) Others	57	123	73	–	–	594
6. Labour (domestic & hired)	**18404 (.567)**	**10860 (.242)**	**6139 (.344)**	**1327 (.397)**	**7683 (.476)**	**44413**
7. Market purchased inputs	**1563 (.048)**	**1432 (.032)**	**459 (.026)**	**40 (.012)**	–	**3494**
Total costs	**27303**	**15821**	**12125**	**1409**	**7683**	**64341**
Net return	**5170**	**29112**	**5732**	**1933**	**8445**	**50392**

Note: Figures in parentheses denote the input-output co-efficient.

CHAPTER NINE

Problems and Constraints

The description of this section is based on multiple responses given by the farmers of the study area. The farmers face many problems, but our analyses is based specifically on the problems of marketing and production process. The information pertaining to these problems are presented in Tables 9.1 to 9.9.

Marketing

Malpractices in the Market

During the survey it was noticed that the malpractices in the market is the main problem. Among these practices, most common were charging high commission, undue charges and unauthorise charges. In Pooh block high commission than usual is main problem according to 76 percent respondents. In Kalpa block multiple charges is the most common problem as revealed by 81 percent respondents. In Nichar block undue charges is a serious problem according to 72 percent farmers.

Transportation

The various problems reported by the respondents like inadequate transport facilities and higher freight charges etc. are presented in Table 9.2. It may be seen from the table that 64.44 percent respondents are facing the acute problem of transportation and 61.25 percent respondents face the problems relating to the higher

---

*See Tables 9.1 to 9.9 at the end of this chapter.

freight charges. It is also reported by 59.38 percent farmers that their villages are not connected by roads. In Pooh and Nichar blocks, lack of vehicle is most serious problem.

Marketing Intelligence

The analysis also revealed (Table 9.2) that 64 percent respondents at overall level complain about late information and about 58 percent receive no information at all about the market. About 59 percent respondents receive misleading information. In Pooh block late information is the main problem, whereas misleading information is the acute problem of the farmers in Kalpa block and no information is the complaint reported by farmers of Nichar block.

Labour Problems

It was seen that there is a problem of labour. The most acute problem in this context is non-availability of skilled labour in all the blocks but 8 percent respondents reported no problem in this regard.

Grading and Packing Material

The problems faced by respondents in various fields of marketing have been presented in Table 9.3. It may be seen from the table that about 78 percent respondents out of the total at overall level are facing the problems related to fruit picking, about 66 percent have problems during grading of fruits, 63 percent have problem of transportation and 72 percent respondents complain about the late payments after the sale of the fruits. About 50 percent experience malpractices in the market.

The most acute problems regarding packing material is concerning the wooden box. About 60 percent respondents complain about other packing material and material not being available. About 54 percent complain about non-availability of credit facilities.

Production Process

Support Prize

About 64 percent respondents out of the total reported that support prices are not announced well in time, while about 49 percent complained that the procurement prices are low in case of potato crop. The glance at individual block reveals that timing of the declaration of support prices is not appropriate.

Seed Certification

The respondents felt that the agency responsible for seed certification is not working properly to meet the local needs of the farmers in the study area (Table 9.1).

Supply of Seed and Planting Material

The problem faced by sampled respondents have been analysed and results presented in Tables 9.5 to 9.7. The results reveal that about 70 percent farmers at overall level faced problems in getting seeds. This problem was reported by about 57, 75 and 76 percent farmers in Pooh, Kalpa and Nichar blocks respectively. About 59 percent farmers reported that they face problem in getting plants from forest department. It is reported by about 76 percent farmers in Pooh, 54 percent in Kalpa and 49 percent farmers in Nichar blocks.

Problems Faced during Migration

A major problem faced by the migrants is of permit for grazing in a particular area, which in H.P. is issued by the State Forest Department. It requires repeated visits and even illegal gratifications, and instances of harassment are very common. At some checkposts taxes are charged from them.

Secondly, the journey during migration is insecure. Sometimes animals are taken away forcibly on the point of guns. No medical facilities are available en route both for men and their animals. During winters they have to take shelter in sheds by paying nominal charges. Sometimes, even after having the permits for grazing, village panchayats assert their rights over the area. They charge substantial amount before letting them graze the animals. The migrants are not provided with any facilities for purchasing from fair price shop and they have to pay more for food as well.

During winters, the animals are sheared at the places of their temporary stay. So, the wool harvested is sold there itself for which itinerant traders visit the places and take full advantage of the poor owners' unawareness regarding prices. All types of tricks are used by the traders in weighment and payment to put the flock owners at a disadvantage. This is a buyers' market in the real sense of the world.

Some of the migrants have purchased their own land in those areas where they migrate during the winter season. They used to shift to that places alongwith family, and cultivate that land during the period of migration.

Other Problems

The people of district Kinnaur are facing different problems like more distance from forest, market and main roads (Table 9.8). Due to the longer distance of forest from the villages people have started substituting fuelwood with LPG or electricity and timber wood with steel and cement. Therefore, people are diverting their attention towards all these modern technology-based facilities (Table 9.9).

Table 9.1. Marketing problems faced by sample farmers

(Multiple response/%)

Block/farm size	Malpractice			Support price		Seed certification	
	More charges	Multiple charges	Undue charges	Not available in time	Less price	Not in time	More formalities
Pooh Block	**76.47**	**58.82**	**49.99**	**51.47**	**35.29**	**35.29**	**41.17**
Marginal	66.66	48.14	37.03	51.85	14.81	25.92	22.22
Small	85.71	60	51.42	45.71	45.71	34.28	45.71
Medium	66.66	100	100	83.33	66.66	83.33	100
Kalpa Block	**61.44**	**80.72**	**65.06**	**59.03**	**55.42**	**79.52**	**55.42**
Marginal	61.29	100	45.16	51.61	22.58	100	9.67
Small	51.35	62.16	67.56	56.75	72.97	62.16	75.67
Medium	86.66	86.66	100	80	80	80	100
Nichar Block	**62.16**	**54.05**	**71.62**	**82.43**	**55.40**	**82.43**	**45.94**
Marginal	100	43.47	78.26	100	52.17	100	21.73
Small	37.5	50	62.5	75	50	75	50
Medium	72.72	90.90	90.90	72.72	81.81	72.72	81.81
All Blocks	**66.22**	**65.33**	**62.66**	**64.44**	**49.33**	**67.11**	**48.00**
Marginal	74.07	66.66	51.85	65.43	28.39	75.31	17.28
Small	57.14	57.14	60.71	59.82	56.25	58.03	57.14
Medium	78.12	90.62	96.87	78.12	78.12	78.12	93.75

Table 9.2. Problems faced by farmers in transportation and marketing intelligence

(Multiple response/%)

Block/farm size	Transportation					Marketing intelligence		
	Lack of vehicle	Not available in time	More freight	Villages not linked with roads	No problems	Late information	No information	Wrong information
Pooh Block	**86.76**	**58.82**	**64.43**	**78.85**	-	**73.53**	**48.53**	**48.52**
Marginal	92.59	25.92	25.22	57.85	-	77.77	48.14	33.33
Small	91.42	85.71	100	100	-	68.67	40	51.42
Medium	33.33	50	33.33	50	-	83.33	100	100
Kalpa Block	**43.37**	**27.71**	**63.85**	**55.41**	-	**66.26**	**60.24**	**73.49**
Marginal	35.48	32.25	48.35	32.25	-	48.38	51.61	64.51
Small	40.54	-	75.67	67.56	-	75.67	67.56	78.37
Medium	66.66	86.66	66.66	73.33	-	80	60	80
Nichar Block	**67.56**	**58.10**	**55.40**	**45.94**	-	**51.35**	**63.51**	**52.70**
Marginal	65.21	82.60	69.56	30.43	-	65.21	69.56	39.13
Small	62.5	37.5	50	50	-	37.5	50	62.5
Medium	90.90	81.81	45.45	63.63	-	72.72	100	45.45
All Blocks	**64.44**	**47.11**	**61.25**	**59.38**	-	**63.55**	**57.77**	**59.11**
Marginal	62.96	44.44	46.67	40.27	-	62.96	55.55	46.91
Small	64.28	40.18	74.11	71.43	-	59.82	52.68	64.28
Medium	68.74	78.12	53.12	65.62	-	78.12	81.25	71.87

Table 9.3. Problems faced by farmers in the field of fruit packing & marketing

(Multiple response/%)

Block/farm size	Picking	Grading	Transport	Payments	Mal-practice	Weather condition	Road condition
Pooh Block	**70.58**	**66.17**	**58.81**	**80.92**	**47.05**	**61.76**	**77.94**
Marginal	70.37	66.66	51.82	85.18	48.14	22.22	51.85
Small	71.42	68.67	62.85	77.14	48.57	91.42	100
Medium	66.66	50	66.66	83.83	33.33	66.66	66.66
Kalpa Block	**71.04**	**59.46**	**68.67**	**50.60**	**54.21**	**59.01**	**61.41**
Marginal	80.6	33.48	96.77	3.22	58.06	70.96	80.6
Small	67.5	75.6	62.16	81.08	54.05	43.2	40.5
Medium	60	73.33	26.66	73.33	46.66	73.33	73.33
Nichar Block	**91.89**	**71.62**	**59.46**	**87.83**	**48.51**	**71.62**	**94.46**
Marginal	100	82.6	47.82	95.62	21.3	86.95	95.22
Small	90	67.5	70	80	62.5	65	95
Medium	81.81	63.63	45.45	100	54.54	63.63	90.90
All Blocks	**77.76**	**65.49**	**62.66**	**72.01**	**50.17**	**63.99**	**77.28**
Marginal	82.70	58.49	67.89	56.78	44.32	59.25	75.17
Small	76.76	70.51	65.18	79.46	55.36	66.05	78.56
Medium	68.75	65.62	40.62	84.47	46.87	68.75	78.12

Table 9.4. Problems faced by farmers in the field of fruit crops

(Multiple response/%)

Block/ farm size	Problems in grading/packing					Problems regarding packing materials				
	Skilled labour	High wages	Not available	No mobility of labour	No problem	Wooden box	Packing material	No credit	Material not available	No problem
Pooh Block	**79.41**	**79.41**	**80.88**	**32.35**	**1.47**	**49.99**	**51.46**	**61.76**	**52.94**	**10.29**
Marginal	81.48	66.66	81.48	22.22	3.70	48.14	29.62	55.55	48.14	25.92
Small	77.14	85.71	77.14	37.14	-	51.42	71.42	65.71	57.14	-
Medium	83.33	100	100	50	-	50	33.33	66.66	50	-
Kalpa Block	**65.06**	**65.05**	**68.67**	**14.46**	**8.43**	**85.54**	**60.24**	**44.93**	**53.01**	**0.00**
Marginal	58.06	80.64	51.61	25.80	12.90	100	45.16	49.35	80.64	-
Small	70.27	48.64	75.67	10.81	5.40	72.97	72.97	24.32	24.32	-
Medium	66.66	73.33	86.66	-	6.66	86.66	60	86.66	66.66	-
Nichar Block	**77.02**	**68.92**	**83.78**	**12.16**	**13.51**	**52.70**	**70.27**	**58.11**	**72.57**	**0.00**
Marginal	82.60	47.82	82.60	39.13	21.73	100	60.86	73.91	78.26	-
Small	80	80	82.5	-	7.5	20	82.5	57.5	75	-
Medium	54.54	72.72	90.90	-	18.18	72.72	45.45	27.27	54.54	-
All Block	**73.33**	**70.66**	**77.33**	**19.11**	**8.00**	**64.00**	**60.88**	**54.35**	**59.55**	**3.11**
Marginal	72.83	66.66	70.37	28.39	12.34	82.71	44.44	58.39	69.13	8.64
Small	75.89	71.42	78.57	15.18	4.46	47.32	75.89	49.10	52.68	0.00
Medium	65.62	78.12	90.62	9.38	9.37	74.99	50.00	62.49	59.37	0.00

Table 9.5. Problems faced by sample farms in various fields

(Multiple response/%)

Block/farm size	Supply of seed		Supply of forest		Supply of fruit nursery		Supply of pesticides	
	Yes	No	Yes	No	Yes	No	Yes	No
Pooh Block	**57.35**	**42.67**	**76.46**	**23.53**	**67.64**	**22.36**	**63.23**	**38.26**
Marginal	74.07	26	88.88	11.11	55.55	30.40	74.07	26
Small	42.85	57.14	74.28	25.71	82.85	17.14	54.28	48.57
Medium	66.66	33.33	33.33	66.66	33.33	16.66	66.66	33.33
Kalpa Block	**74.69**	**32.52**	**54.21**	**46.98**	**37.27**	**61.44**	**60.23**	**39.76**
Marginal	83.87	32.25	64.51	38.70	19.35	77.41	58.06	41.93
Small	67.56	35.13	54.05	45.94	56.75	43.24	59.45	40.54
Medium	73.33	26.66	33.33	66.66	26.26	73.33	66.66	33.33
Nichar. Block	**75.67**	**25.67**	**48.65**	**50.00**	**58.11**	**43.24**	**67.56**	**32.43**
Marginal	86.95	17.39	39.13	56.52	56.52	47.82	82.60	17.39
Small	80	20	60	40	55	45	62.5	37.5
Medium	36.36	63.63	27.27	72.72	72.72	27.27	54.54	45.45
All Block	**69.77**	**33.34**	**59.11**	**40.88**	**53.30**	**43.64**	**63.55**	**36.90**
Marginal	81.48	25.95	65.43	34.56	41.97	53.34	70.36	29.65
Small	64.28	36.60	62.50	37.50	64.28	35.71	58.92	41.96
Medium	59.37	40.62	31.25	68.74	43.56	46.87	62.49	37.50

Table 9.6. Other problems faced by sample farms in the field of irrigation and farm machinery

(Multiple response/%)

Block/farm size	Supply of weedicides		In the field of irrigation		Farm machinery	
	Yes	No	Yes	No	Yes	No
Pooh Block	**58.82**	**29.48**	**79.37**	**23.53**	**51.47**	**41.17**
Marginal	55.55	15	85.18	22.22	62.96	22.22
Small	68.67	31.42	74.20	25.71	45.72	51.42
Medium	16.66	83.33	83.33	16.66	33.33	66.66
Kalpa Block	**51.30**	**48.18**	**52.67**	**47.95**	**46.98**	**54.21**
Marginal	60.29	38.70	30.70	70.96	16.12	83.87
Small	40.24	59.45	60.00	40.00	62.16	37.83
Medium	60.00	40	80	20	73.33	33.33
Nichar Block	**44.50**	**54.07**	**70.27**	**29.73**	**52.70**	**45.92**
Marginal	60.56	39.18	100	0	73.91	26.00
Small	40	57.5	60.0	40	42.5	55
Medium	27.27	72.72	45.45	54.54	45.45	54.54
All Blocks	**51.34**	**44.47**	**66.53**	**34.58**	**50.22**	**47.54**
Marginal	58.79	30.94	68.54	34.56	48.14	46.89
Small	49.01	49.99	64.44	35.53	50.00	48.21
Medium	40.62	59.37	68.75	31.25	56.25	46.87

Table 9.7. Other problems faced by farmers

(Multiple response/%)

Block/farm size	Electricity/Diesel			
	No electricity	Low voltage	Less supply	Operational difficulties
Pooh Block	**14.71**	**72.05**	**66.17**	**38.23**
Marginal	11.11	81.48	66.66	33.33
Small	20	71.42	68.57	42.85
Medium	-	33.33	50	33.33
Kalpa Block	**15.66**	**66.26**	**75.90**	**49.39**
Marginal	22.58	48.38	80.64	48.38
Small	16.21	72.97	78.37	35.13
Medium	-	86.66	60	86.66
Nichar Block	**9.46**	**51.35**	**63.51**	**58.11**
Marginal	13.04	69.56	73.91	39.13
Small	10	40	55	60
Medium	-	54.54	72.72	90.90
All Blocks	**13.33**	**63.11**	**68.89**	**48.88**
Marginal	16.05	65.43	74.07	40.74
Small	15.18	60.71	66.96	46.42
Medium	0.00	65.62	62.50	78.12

Table 9.8. Average distance from forest, market, road, kuhl and own land of sample households

(Km.)

Block/farm size	Destination				
	Forest	Market	Road	Kuhl	Own land
Pooh Block	**4.31**	**13.99**	**3.63**	**2.69**	**7.87**
Marginal	4.51	11.55	5.29	1.66	4.91
Small	3.7	15.71	2.85	3.37	9.59
Medium	7	14.88	1.66	3.33	11.15
Kalpa Block	**1.59**	**6.84**	**2.37**	**2.53**	**8.18**
Marginal	1.38	6.9	1.58	1.62	5.98
Small	1.62	7.83	3.32	2.91	8.7
Medium	1.93	4.26	1.66	3.46	11.46
Nichar Block	**2.71**	**5.90**	**2.98**	**0.75**	**7.86**
Marginal	1.78	3.6	1.89	0.43	6.33
Small	3.35	6.62	3.77	0.87	7.96
Medium	2.36	8.09	2.36	1	10.72
All Blocks	**2.78**	**8.69**	**2.95**	**1.99**	**7.98**
Marginal	2.54	7.51	2.90	1.30	5.72
Small	2.89	9.86	3.33	2.33	8.71
Medium	3.03	7.57	1.71	2.59	11.15

Table 9.9. Substitute for fuelwood/timberwood being used by sample farms

(Multiple response/%)

Block/farm size	Substitute			
	Cement & iron	LPG	Electricity	Solar/biogas
Pooh Block	**51.46**	**92.34**	**60.29**	**89.7**
Marginal	33.33	80.7	74.07	100
Small	71.42	100	48.57	82.85
Medium	16.66	100	66.66	83.33
Kalpa Block	**26.50**	**88.19**	**96.39**	**79.52**
Marginal	16.12	68.38	100	96.77
Small	40.54	100	100	70.27
Medium	13.3	100	80	66.66
Nichar Block	**32.43**	**87.66**	**97.30**	**70.27**
Marginal	34.78	60.30	100	47.82
Small	32.5	100	100	100
Medium	27.27	100	81.81	9.09
All Blocks	**35.99**	**89.27**	**85.78**	**79.55**
Marginal	27.16	70.19	91.36	83.95
Small	47.32	100.00	83.93	84.82
Medium	18.73	100.00	78.12	50.00

CHAPTER TEN

Summary, Conclusions and Suggestions

A farming system is a complex human activity. It consists of different farm and off-farm activities which are inter-dependent and interact for available resources within a socio-political environment. Agriculture, horticulture, livestock and farm land forestry are the major sub-sectors of the farming system. Common property resources (CPRs) system acts as supplementary inputs for the existence and operations of different farming systems. This investigation was carried out in Kinnaur district of Himachal Pradesh to study the economics of various farming systems, contribution of CPRs and linkages between different components of farming systems.

This study specifically focuses on the following objectives: (i) to identify the alternative systems of farming in dry temperate zone of H.P.; (ii) to study existing farming systems and their inter-linkages in Kinnaur district of Himachal Pradesh; (iii) to analyse the depen-dence of farm enterprises of various farming systems on the local common property resources (CPRs) and; (iv) to analyse the policy imperatives of existing farming systems in the study area.

Research Methodology

Most of the literature on farming systems includes investigations and comments. There is evidence of comprehensive research on the study of farming systems, but it is realised that some fields still remain unexplored.

Informations were collected through schedules which were administered on the sample of 255 farmers (out of this 225 farmers provided complete information). Multi-stage random sampling was used during the survey. The survey was conducted for the year 1997-98. All the three blocks: Pooh, Kalpa and Nichar of Kinnaur district were taken for the study for the proper re-presentation of Kinnaur district.

Both primary as well as secondary data were collected for the various objectives of this study. Primary data regarding socio-economic features of household, agricultural crops, horticultural crops, livestock, farm forestry, common property resources, management practices etc. were collected by personal survey method on specifically prepared and pre-tested schedules. The secondary data for the study were collected from different published and unpublished documents obtained from various offices and personnel.

Tabular method, involving percentages and averages has been used to study the socio-economic features, economics of existing farming systems and contribution of CPRs.

To study the linkages between the various farm components of various farming systems and common property resources static input-output model has been used. Other statistical tools were also used for analysis and interpretation of the data.

Conclusions

The conclusions of the study are grouped as under: (a) main conclusions based on secondary data, and (b) main conclusions based on primary data.

Main Conclusions based on Secondary Data

Secondary data are used to analyse changes in land use pattern, cropping pattern, production and productivity of major crops, demographic features, level of literacy, distribution of worker and composition of livestock. Secondary data-based findings for the study are as follows:

1. The analysis of land use pattern in district Kinnaur reveals that share of forest in reporting area is recorded 22.34 (000 ha.) out of 640.1 (000 ha.), which is less than 5%. The total geographical area of tribal region in the state is 23,655 sq. km. of which permanent pastures and grazing land account for 65.46 percent. Net area sown forms only 1.72 percent of the total geographical area in Kinnaur which is lesser than the state figures of 20.83 percent for the year 1992-93. These facts inter alia, the higher proportion of workers depending on agricultural pursuits, suggests that it is not cultivation of crops alone on which the people of Kinnaur district should generally be depending. The area sown more than once is also less than the state figures. Therefore, the cropping intensity in study area is lower ((123 percent) than the state average (170 percent). The low cropping intensity in Kinnaur district is due to the generally prevalent inclement weather in the area. Other use of land such as fallow lands and area not available for cultivation etc. is negligible. Alpine pastures contributed the largest area of about 45 percent of the total geographical area of Kinnaur district

Overall picture of the cropping pattern in Kinnaur shows that the fruits, vegetables and pulses (rajmash) are the most important crops and have a great scope of development in this particular area. But due to absence of needed infrastructural facilities like transportation storage, processing etc. the potential could not be tapped. Furthermore, the proportion of marketable surplus in fruits and vegetables is much higher than other crops. Horticulture is the fastest growing activity in this area. The area is most suitable for growing off-season vegetables.

2. The analysis of demographic features in Kinnaur district reveals that the geographical area of Kinnaur is 6401 sq. km. Population is 71,270 and density of population is 11 against the density of H.P. (93). Growth rate of population of study area is 19.7% and there is no urban population is Kinnaur district. Literacy

of Kinnaur district in 58.4% out of which 72 percent male and 42 percent female are literate. Out of the two Govt. degree colleges running in tribal area, one is in Kinnaur (Recong Peo).

3. Demographic factors affect the potential of labour force of any area, Women have always played an important part in the labour market. This is most true for the study area where majority of people depend on horticulture and agriculture; it is the females who actually bear the burden of all the work except ploughing. The number of females per thousand stands at 856 in Kinnaur which is lower than the state (976).

4. The economy of an area depends much on the strength of active workers also. In 1991 Census most workers comprised cultivators, agricultural labours, those involved in household industries and other workers including factory workers, plantation workers, those in trade and commerce, business, transport, mining, construction and all Government and Municipal employees. Another category of workers i.e. marginal workers have also been introduced to provide an estimate of under-employed persons. Thus the total active labour force will be the sum of main workers and marginal workers which is about 90.26% out of the total workers' population. The main worker who have worked for six months or more in a year are taken as fully employed . The extent of under-employed in Kinnaur is less than 5 percent of the total population. However, in actual practice all the main workers can not be taken as fully employed. Therefore, the extent of under-employed will definitely be higher than the estimate based on marginal workers only.

5. The importance of livestock to the state economy of Himachal Pradesh has been significant. In Himachal Pradesh sheep and goats are generally owned by nomadic and semi-nomadic people, who belong to tribal region of the state. The study area accounts maximum of sheep and goats in the livestock. Sheep and goat farming is a hereditary occupation with the people of Kinnaur district. The high mountain pastures provide excellent grazing facilities during summer months. Due to heavy snowfall

during winter, the flock masters have to take their flocks to far-off places in the lower hills. Change in the livestock in Himachal Pradesh during 1972 to 1992 shows that there is maximum growth in the equine (horses, ponies, mules & donkeys) population. The same is noticed in Kinnaur district. Sheep and goats and improved quality of cows, generally have been given special attention by government. But goat population is again discouraged by the government because they cause maximum damage to the vegetative cover and the soil resulted in erosion. In ovine category the number of sheep is dominating in tribal region. It is observed that nearly 68 percent goats are reared in Kinnaur district.

6. Yak is the most important animal for the people living in Kinnaur district. It is the only species of animal which produces milk, meat, hide, wool and is also used for transportation at higher altitudes where no other ruminant can thrive economically. The yak can dig out roots of vegetation even buried under snow as deep as 10-15 cm. It is high climber and can plough through snow in winter. The functional comparison of yaks with the camels can be safety termed as ship of the highland or camel of the cold desert. Yak meat is considered to be quite delicious. Yak hair is generally used to make mattresses, shawls, rugs, ropes, tents, and the bushy tail is used as 'chaunri' in religious places. In an agro-pastoral economy like that of Kinnaur where the mechanism of agricultural operations is not possible, livestock are kept not only for providing protective food such as milk, meat, etc., but also for the valuable drought power and manure essential for agriculture in the region.

From the above discussion, it may be concluded that the potential of livestock farming in Kinnaur district is very high.

Main Conclusion based on Primary Data

For the scope of farming system and their impact over district Kinnaur is studied by using primary data and findings of this study are as under:

1. Average family size of the selected households in different blocks of Kinnaur district is 6.67. Family size has direct relationship with the farm size. The joint family system is noticed highest in medium farm category. Polyandry system, which was prevalent earlier is now declining with the spread of education.

2. The literacy percentage in the study areas is highest in Nichar block (88.19%) and lowest in Kalpa (85.85%). The literacy rate among female is encouraging i.e. 83.46 percent out of which 27.28% were matriculate, 8.41 percent graduate and only 2.02 percent are found postgraduate.

3. The boundary of Kinnaur in eastern side touches western Tibet. The impact of Buddhism can be seen very easily. Almost every village has gompa (Buddhist temple). In the whole district of Kinnaur 80.44 percent belongs to Buddhism and 19.56 percent to Hinduism.

4. The number of active workers per bigha of operational land and gross cropped area decreases with the increase in the size of holding. The maximum working force is engaged in horticulture followed by agriculture, while off-farm activity is the other important source of livelihood. The total value of farm implements and machines are almost equal in every blocks. Human labour shares the maximum portion of total cost.

5. Total land holding in all the three blocks ranges from 8.81 to 68.66 bighas out of which more than 50 percent is under field crops. Lot of inequality is observed in the ownership pattern of land holdings. Marginal farm groups some time lease the land from medium farms. The assets held by the households show quite glaring inequalities.

6. Livestock is an important component of dry temperate zone farming system. It is dependent on forest which make livestock system viable. The livestock system in this area consists of few but improved quality animals for milk and cow dung purposes. Now people have paid attention towards stall- feeding. The number of improved breed animals increases with the farm size. Livestock

is the only source of drought power in the area, besides it provides milk, meat, wool, hides and skins to satisfy the need of farm households. Animals are also used for transporting the products and inputs. On an average 1.18 cows in milk, 1.24 cows dry, 1.30 bullocks, 1 yak, 0.51 horses/mules, 10.16 goats and 16.37 sheep have been found per household. Ovines (sheep and goats) are most profitable component in livestock. But to look after ovines requires hard work. Area is snow bound, demand for fodder increases. So, people migrate their livestock specially ovines and equines to lower hills during winter for grazing. The average period for migration is noticed as 4.24 months during winter.

7. Number of fruit trees planted on the farms is lowest in Pooh and highest in Nichar block. Apple is the main fruit tree of the study area. The average holding under apple is 37 percent of the total cropped area. The proportion of trees in bearing stage is found to be almost equal in all three blocks. People are growing other fruits like almond, apricot and peas. Initial cost of plantation of apple was found on an average at Rs. 36. The marketing expenditure per box, from picking in orchard to different distant market is found to be Rs. 120 in district Kinnaur. People of this area also have complaints about malpractices in market, labour problem and lack of marketing intelligence. Return wise, in fruit crops Pooh is first (per household Rs. 33,483) followed by Kalpa (Rs. 28,458).

8. Number of forest trees are found negligible in this area. Higher emphasis of farmers on fruit cultivation is the reason for low forest tree inventory of sampled farms. Maximum forest trees in the area are in the sapling stage.

9. Wheat barley, maize, rajmash, peas and potato are the major crops accounting for significant proportion of the cropped area. In Pooh and Kalpa block almost 90 percent land is irrigated. The proportion of wheat, barley, maize, rajmash and potato in gross cropped area is worked out to be nearly 16 percent, 14 percent, 8 percent, 8 percent and 4 percent respectively. In food crops rajmash is realised highest return-wise. The farmers' attraction towards the HYV seeds is noticeworthy. It is observed that farmers

are concentrating more on horticulture than agriculture which according to them is loss incurring. Crop rotation in this area is usually single crop based due to long winter season. So, cropping intensity in this zone is very less.

10. Working capital requirement for horticulture is highest. The total requirement of human labour is highest under Nichar block (755 mandays) followed by Pooh block (612 mandays). The labour requirement of medium farm category is about 40 percent higher as compared to small farm at overall level. The major proportion of the total labour is used in livestock component under different farming sub-systems followed by horticulture and CPRs. The farmers generally use hired labour for conducting various farm operations. Peak labour demand is in September onward upto November and labour demand exceeds the labour supply. In other months labour availability is more than demand. During slack period some farmers enjoy leisure time while others perform non-farm activities. It was also observed that labour requirement for livestock and CPRs remains more or less evenly distributed throughout the whole year. Farming system in this area can be categorized as remittance-dominated mixed farming system.

11. Off-farm activities have emerged as the most important source of family income in this area and the same have contributed about 14.42 percent of the total household income. Maximum contribution of off-farm income is from government job followed by pension, trade/business and horticultural labour respectively. A further break up reveals that people of this area are diverting towards secondary and tertiary activities.

12. Off-season vegetable cultivation also has emerged as the most important farm activity in the study area. The main vegetables grown are tomato, capsicum, peas, chilly, cabbage, cauliflower and potato. It is again noticed that there is no crop rotation in vegetable crops. The vegetables other than potato and peas production is only for self-consumption. But peas and potato are most capital-intensive crops. Pea is emerging as major cash crop especially in Pooh block. For the welfare of people Govt. Vegetable

Research Centre is running at Kalpa. But People are more concerned with horticultural crops, otherwise, farming system of this zone can be vegetable-dominated mixed farming system.

13. Fruit cultivation has emerged as the most important farm activity, as it contributed 49.44 percent of the total net household income followed by CPRs 14.34 percent, agriculture 9.73 percent and livestock rearing 8.78 percent. Crop income per bigha is Rs. 1701 at overall level. It is further observed that per capita income in Kinnaur district is Rs. 8828 which is below the H.P. level. Per capita income is highest in Pooh block (Rs. 11,429) followed by Kalpa block (Rs. 8646).

14. Fuelwood requirement is more due to long winter season. The requirement is fulfilled from CPRs, farm forestry and agriculture/horticulture waste. Average fuelwood consumption is 30.61 qtls. farm/year. Further break up reveals that people are using LPG, kerosene oil, electricity and coal as an emerging source.

15. Commercial banks are the main source of credit. Generally credit is required for plantation of fruit trees, renovation of land, marketing of commercial crops up to distant markets and purchasing of households etc. During field survey it is noticed that people have no complaints about the credit facilities.

16. It is revealed that crop husbandry, livestock husbandry and forestry constitute the main and closely integrated components of farming systems. The farmers who cultivate land also raise livestock and depend on the forestry for support of both the farming components. A change in any one of the component of farming system significantly affects the other and results into breakdown of the system. Important linkage exists between various components of farming. In the process of production there is employment creation as well as income generation.

17. Over centuries, the forest and other common property resources have formed an important part of the Kinnaur district of Himachal Pradesh. A strange relationship has always existed between forestry and life, culture and economy of the great

multitude of rural populace. Kinnaur district meet out the requirements of fuelwood, timber, fodder, litter, grazing, fibres and fruits etc. either from own private forest land or from public or from both. However, the nature and degree of dependence varies according to purpose, income status of people and agro-climatic conditions. In fact, over a greater part of hilly areas, sheep and goats are mostly dependent on grazing. In addition to grazing there is a massive supply of cut-fodder from forest areas in all weather and particularly in time of scarcity. Farm land forestry provides agricultural implements, poles for livestock sheds, fuelwood, staking material, grass and grazing, litter etc. The government forests in this area provide timber and poles for livestock sheds, housing, furniture, renewed regularly to sustain the hill farming practices. Among the other forest products derived from the forest are livestock fodder for stall-feeding, grazing etc. Thus, the community property resources (CPRs)-farming system inter-relationship is crucial to the existence of Kinnaur farmers.

18. Actual consumption expenditure and viability reveals that all the farm categories are almost viable in Kinnaur district. Further break up indicates that non-food item expenditure is highest among all items. This unproductive expenditure needs curbing and credit should be given towards production purposes. As a whole in Kinnaur district income from all sources is Rs. 58,885 per farm/ year against the total expenditure of Rs. 50,365, thus getting surplus of Rs. 8520. It indicates that all farm categories in Kinnaur district are getting surplus after meeting all expenditure during the year. It may, therefore, be concluded that there is a need for enhancing the farming income in this district, so that in addition to meeting out the consumption expenditure, the farmers are able to save for further investments on the farm which are required for improving income generation and thus, standard of living of the farming families.

19. Common property resources are the resources accessible to the whole community of a village and to which no individual has exclusive property rights. In Indian context, this category of

resources includes community pastures, common lands, forests, panchayat forests, threshing grounds, rivers, common paths, kuhls, temples etc.

20. Forests, village common lands, kuhls and panchayat forests are identified as the major CPRs in study area. The inhabitants of this area are having the customary rights for grazing, grasses, fuelwood, timber and wood for implements. The grazing of livestock is a traditional practice in Kinnaur district. The extent of grazing is highest for ovines (sheep & goats) followed by bullocks and young stocks. Total demand of fodder and concentrate is completed from CPRs, farm resources and markets. Total annual contribution from CPRs is estimated at Rs. 14,692, Rs. 17,574, Rs. 15,826 and Rs. 16,128 in Pooh, Kalpa, Nichar block and at overall level respectively.

21. The allotment of the forest lands for the agricultural purposes, grazing in the planted area, felling of green plants have been banned by the forest department in light of 1980 Forest Conservation Act and under new forest policy of 1988. The other rights also have been curtailed. Animal grazing, illicit felling, lopping of green tree leaves are the major threats identified by forest department. Beneficiaries have not participated in the management of CPRs except for fire fighting in study area.

22. It is revealed that production potential of CPRs has declined considerably during the recent times and arrangements for the management of CPRs are inadequate in Kinnaur district. It is found that the condition of village common lands and panchayat forests are poorest in terms of management. Lack of adequate manpower, funds, cooperation of users' group in the management and rehabilitation programme are main reasons for this decline.

23. 'Kuhls' the only source of irrigation in district Kinnaur is managed and maintained under common property regime. The contribution of kuhl in agriculture and horticulture in the study area is very high. The average area under irrigation at overall level in Kinnaur district is found 62 percent of the cropped area. The

use of kuhl is highest in Pooh block followed by Kalpa block.

24. Kuhls are not in a good condition in this area. Because government has not given proper attention towards irrigation facilities. The people's joint management is working, but is unable to manage it properly.

25. Main operations in kuhl irrigation system include 'Danga' repair, water distribution and conflict management. Among above said operations 'Danga' is considered very important. Joint management committee of villagers used to repair it with the help of other persons. The activities like water theft are supervised by the village committee.

Suggestions

From the above findings the requirement of societal and other environmental factors, the following remedial measures may be taken for the improvement of the farmers of the study area. The suggestions of the study are grouped as under: (a) suggestions regarding improvement of horticulture and agriculture, (b) suggestions regarding better management of CPRs, and (c) suggestions regarding improvement/management of kuhls.

Suggestions Regarding Improvement of Horticulture and Agriculture

1. Credit facilities which are not sufficient for the farmers should be provided more extensively. For this purpose wide network of banks should be created in rural and far-flung areas of Kinnaur district.

2. Poor transport network alongwith distantly located markets are the major hurdle in agricultural and horticultural development. So, all weather roads should be developed.

3. Timely availability of various inputs is necessary for the successful agricultural and horticultural development.

4. Cooperative marketing should be encouraged for the farm surplus.

5. Off-season vegetables should be grown in this area in large scale. Because vegetable seeds produced in Kinnaur district are vigorous and disease-free.

6. Bee-keeping is necessitated for the pollination services. Migratory bee-keeping should be popularized in this area.

7. For the diversification of farm income, horticulture and forestry plantation may be done on the land not suitable for field crops.

8. Efforts should be made to shift surplus labour (disguised unemployment) from farm to off-farm activities.

9. The climate of some pockets of the study area is more suitable for floriculture, specially for the cultivation of gladiolus, pansy and carnation.

10. The subsidy in Kinnaur district is enjoyed by every farm category almost at the same rates. But it should be diverted towards the marginal farmers.

11. More seminars, camps and exhibitions should be organized for the awakening of the farmers.

12. The climate of Kinnaur district is suitable for energetic dry fruits like almond, apricot etc. So, more stress should be given to the production of dry fruits.

Suggestions Regarding Management of CPRs

1. The farmers of the study area are dependent on common property resources (CPRs). Efforts should be made towards their protection, regeneration and development.

2. Encroachment and other illegal activities in the CPRs should be checked.

3. An area in a pasture should be earmarked and closed for production of seeds of desirable grasses.

4. Provision for grasses from the particular enclosures for particular month in a year should be made.

5. The rotation/different grazing systems may be followed to give the advantage of early nutritive growth.

6. The activities of mining/quarrying should be stopped immediately.

7. The timber distribution rights (T.D.R.) are generally more than their genuine need in some cases. So, T.D. should be regulated only for genuine needs.

8. Due to cold desert region, there is an acute shortage of fuel and timber and people depend largely on forests. So, people should be encouraged to use L.P.G. biogas and solar energy etc.

9. The multipurpose tree species should be preferred in this area.

10. The varieties of grasses can also be incorporated in the degraded forest sites and common lands.

11. People should be educated to keep improved animal breeds under stall-feeding.

12. Scientific exploitation of minor forest products should be encouraged to supplement the income of the people.

13. The chilgoza forest, being a source of income to the local people should be put under effective management.

14. The region is endowed with rich fauna and should be carefully preserved.

15. Undesirable growth of weeds and unpalatable grass species should be eradicated manually or by the use of weedicides to obtain high production level.

16. It is realised that due to the construction of roads and projects, the debris have largely gone into stream or other open places. The curative measures should be adopted.

17. All small slides should immediately be given corrective measures to prevent a serious problem.

18. Common paths, temples, underground water resources which are used commonly should be maintained properly.

19. People's involvement can be increased by linking the benefits with the management responsibilities under the joint management of CPRs.

Suggestions Regarding 'Kuhl' Improvement/Management

1. The irrigation channels must be *pucca* or lined to avoid seepage to the nearby areas prone to landslides.

2. Besides tapping more surface water the lift irrigation should be made functional to have more irrigation.

3. Kuhl committees should be strengthened. So that rules can be implemented strictly.

4. If a member of kuhl cannot participate in any of kuhl operation, he should deposit the equivalent amount with the committee to hire labour well in time.

5. Financial aid from government for repairs and development works may be invested on the direction of kuhl committee.

6. A joint account from the contribution of members should be raised for the unforeseen repairs of kuhls.

7. Water user's right should be maintained as per the revenue records and water distribution should be maintained as per the traditions.

8. Minor unforeseen repairs should be done by the water users under the control of kuhl committee. The repair & maintenance and other departmental works are required to be completed every year.

9. Government intervention in kuhl committee should not be biased.

10. Irrigation, public health department and other government agencies should pay more attention towards irrigation facilities.

Bibliography

Agarwal, A.K., 1988. "Forestry Development in North East: An Approach". Advances in Forestry Research in India, 105-116.

Agarwal, G.D. and S.M. Pathak, 1957. "Budgeting Approach in Farm Planning". Agriculture Situation in India, 12(3): 238-248.

Anonymous, 1985. Draft Seventh Plan, 1985/90 and Draft Annual Plan 1985/86. Shimla: Planning Department, Govt. of H.P.

Baker, M., 1995. "The Gravity Flow of Irrigation System of Kangra Valley of Himachal Pradesh". Contemporary Transformations and Interactions with Farming Systems. Paper presented in International Seminar on Natural Resource Management and their Linkages with Farming Systems, June 8-9, UHF, Nauni-Solan (H.P.)

Bhati, J.P., 1981. Population, Resource Use and Ecological Stresses in the Western Himalayas. Ph.D., Dissertation. University of Hawaii, Honolulu.

Bhati, J.P., 1983. Challenges for Rural Development in Hilly Areas—A Case of Himachal Pradesh. Cooperative Perspective (Pune, India), 18(3): 101-117.

Bhati, J.P., 1990. Development Strategies in Himachal Pradesh. Mountain Farming Systems. Discussion Paper No. 6. Kathmandu: International Centre for Integrated Mountain Development.

Bhati, J.P. et al., 1992. "Diversity of Mountain Farming Systems in Himachal Pradesh Himalaya". Chapter in N.S. Jodha (ed.). Sustainable Mountain Agriculture, New Delhi: Oxford and IBH Publishing Company.

Bhati, J.P. and C. Gopalakrishnan, 1979. "Identifying Optimal Farming Systems for the Western Himalayan Region: An Ecological Approach". Indian Journal of Agricultural Economics, Vol. XXXIII (4): 56-57.

Bhati, J.P. and L.R. Sharma, 1996. "Farming System Approach to Watershed Planning for Development of Hilly Areas: A Study of Himachal Pradesh". In S.L. Shah (ed.), Agricultural Development in Hilly Areas: Constraints and Potential, New Delhi: New Agricultural International Publisher.

Bhati, J.P. and L.R. Sharma, 1996. "Forests, Agro-Forestry and People in Western Himalayas". Ecology, Vol. 11, No. 1.

Bhati, J.P. and R. Swarup, 1985. "Why Eco-development Fails in the Himalayas". In T.V. Singh, and J. Kaur, (eds.), Integrated Mountain Development. New Delhi: Himalayan Books, 337-348.

Bhati, J.P. and W. Peter Zingel, 1987. "Rural Development and Environmental Management in Hilly Areas". In Srivastava, M. and A.K. Singh, (eds.), Rural Development in India: Approaches, Strategies and Programmes. New Delhi: Deep and Deep Publications, 97-108.

Bist, N.S., 1984. "Strategy for Integrated Agricultural Development of Uttarkhand Himalaya". Rural Development in the Himalya Problems and Prospects, 152-160. Gyanodya Prakashan, Nainital.

Bromley, D.W. and M.M. Cernea, 1989. The Management of Common Property Resources: Some Conceptual and Operational Fallacies. World Bank Discussion Papers: 57-66.

Chand, R., 1994. "Role of Water Rights in Farmer-Managed Hill Irrigation Systems". Economic and Political Weekly. 29(13): A26-A30.

Chatterjee, B.N., 1992. "Prying Open the Poverty Trap with Complete Farming Systems Approach". Indian Farming, 42(1): 10-16.

Chopra, K. and G.K. Kadekodi, 1991. "Participatory Institutions: The Contest of Common and Private Property Resources". Environmental and Resource Economics. 1(4): 353-372.

Chopra, K., G.K. Kadekodi and M.N. Moorty, 1990. "Participatory Development—People and Common Property Resources". Sage Publications, New Delhi, 163pp.

Damodaran, A., 1988. "Morphology of Grazing and its Crisis in Sedentary Communities". Economic and Political Weekly. 23: A29-A34.

Daniel, E.F., 1960. "Reports on Farm Management and Planning". Government of India, New Delhi.

Duckham, D.N. and G.B. Masefield, 1970. Farming Systems of the World. New Delhi: Allied Publishers.

Farrington, J. and A. Martin, 1987. "Farmer Participatory Research: A Review of Concepts and Practices". In: ODI Agricultural Administration (Research and Extension) Network, Discussion Paper No. 19.

Fischer, A.C. and J. Krutilla, 1974. "Managing the Public Lands: Assessment of Property Rights and Valuation of Resources". Haefele: 35-39.

Gangopadhyay, S., 1991. "Agricultural Characteristics in Agro-Climatic Zones of Himachal Pradesh". In Agro-Climatic Zone Specific Research ed. by S.P. Ghosh. New Delhi, ICAR: 158-172.

Garen, J., 1993. "Aspects of Common Property Resource use in India". TRI News Spring: 18-19.

Gibbs, C.J.N., D.W. Bromley and F. Berkes, 1989. "Institutional Arrangements for Management of Rural Resources—Common Property

Regimes". In Ecology and Community Based Sustainable Development (ed. Berkes, F): 22-32.

Gilmour, D.A. and M.C. Nurse, 1991. "Farming Initiatives in Increasing Tree Cover in Central Nepal". Mountain Research and Development. 11(4): 329-337.

Gupta, A.K., 1986. "Socio-Ecology of Grazing Land Management". Proceedings of Second International Range land Congress, Adelaide, Australia. 13-18 May 1984. (eds. Joss, J.P., P.W. Lynch and D. B. Williams): 289-291.

Gupta, R. and et al., 1992. " Status of Fuelwood and Fodder Balance and its Impact on Farm Woman in Himachal Pradesh and Jammu & Kashmir". Regional Centre, NAEB, UHF, Nauni-Solan: 80pp.

Gupta, T., and S. Sambrani, 1978. "Control of Shifting Cultivation: The Need for the Integrative Approach and Systematic Appraisal". Indian Journal of Agricultural Economics, Vol. XXXIII(4): 1-18.

Hartwick, J.M. and N.C. Olewiler, 1986. "The Economics of Natural Resource Use". Harper Collins Publishers: 527pp.

Ives, J.D. and B. Messerli, 1989. "The Himalayan Dilemma: Reconciling Development and Conservation". London, Routledge.

Jodha, N.S., 1986a. "Research and Technology for Dry Land Farming in India—Some Issues for Future Strategy". Indian Journal of Agricultural Economics, 47(3): 234-247.

Jodha, N.S., 1986b. "Common Property Resources and Rural Poor in Dry Region of India." Economic and Political Weekly, 30(27): 1169-1181.

Jodha, N.S., 1989a. "A Framework for Integrated Mountain Development, Mountain Farming Systems". Discussion Paper No. 1, Kathmandu: International Centre for Integrated Mountain Development.

Jodha, N.S., 1979. "Inter-cropping in Traditional Farming Systems". ICRISAT, Economics Progress Report-3, Patancheru, A.P., India.

Jodha, N.S., 1989b. Mountain Farming Systems: Search for Sustainability, Mountain Farming Systems. Discussion Paper No. 2. Kathmandu: International Centre for Integrated Mountain Development.

Jodha N.S., 1990. Mountain Perspective and Sustainability: A Frame Work for Development Strategies. Paper Prepared for International Symposium on Strategies for Sustainable Mountain Agriculture, Kathmandu: ICIMOD, 10-14 September.

Joshi, P.K. and N.K. Tyagi, 1991. "Sustainability of Existing Farming System in Punjab & Haryana—Some Issues on Ground Water Use." Indian Journal of Agricultural Economics. 46(3): 412-421.

Kahlon, A.S. and S.S. Johl, 1967. "Application of Programming Techniques to Indian Farming Conditions." Ludhiana, Department of Economics and

Sociology, Punjab Agricultural University.

Krishnamurthy, K. S., Purushotham and M.K. Munegauda, 1987. "Relevance to Silvi-Pastural Agro-Forestry and Range Land Farming Systems—Concepts". UAS Technical Series, University of Agricultural Sciences, India. 47(11): 28-32.

Lipton, Martin., 1997. "Farming Systems in Economic Development". Economics of Sub-Tropical Farming System (low priced ed.). Cambridge University Press: U.K. (Great Britain): 1-23.

Maghembe, J.A., T.M. Chingaipe and L.L.L. Lulandala,1981. "Agro-Forestry and Farming Systems: Role in Marginal Agricultural Areas". Proceeding of the Conference on Farming Systems and Farming Systems Research, held at the International Conference Centre, Arusha, Tanzania, April 14-16: 96-103.

Mahat, T.B.S. 1986. "Forestry Contribution to Upland Farming Economics in the Hindu-Kush Himalayan Region." ICIMOD Working Paper. International Centre for Integrated Mountain Development, Kathmandu.

Mahat, T.B.S., 1987. Farming Forestry Linkages in the Mountains. ICIMOD Occasional Paper No. 7, Kathmandu: International Centre For Integrated Mountain Development.

Moorti, T.V., R.C. Oberoi and D.R. Thakur, 1990. "Impact of Sheep and Goats on the Economy and Environment of High Altitude Areas of Himachal Pradesh". Final Technical Report. Department of Agricultural Economics, HPKV, Palampur: 213 pp.

Nayer, 1985. "A Talk on Agroforestry Systems". ICAR Research Complex for NAEB Region. News Letter, 8(4): 7-8.

Norman, D.W.C., 1978. "Farming System Research to Improve the Livelihood of Small Farmers". American Journal of Agricultural Economics, 60(5): 813-818.

Norman, D. and M. Collinson, 1985. "Farming Systems Research in Theory and Pratice". Proc. ACIAR Workshop on Agricultural Systems Research For Developing Countries held at Hawkesbury Agril College, Richmond, Australia, May 12-15: 16-30.

Oberoi, R. and T.V. Moorty, 1986. "Economic Analysis of Sheep Farming in Tribal Farms of Bharmaur Tehsil", Himachal Journal of Agricultural Research, 12(1): 35-46.

Oommen, M.A. 1963. "The Economics of Cropping Pattern—A Case Study of Kerala". Indian Journal of Agricultural Economics, 18(1): 120-128.

Pandey, R.N. and T.S. Bhogal. Approach to the Eighth Five Years Plan 1990-95, New Delhi: Government of India.

Parasnis, S.S., 1976. Forestry In The Rural Economy of India, with Particular Reference to Agriculture. Ph.D. Dissertation, The University of

Michigan, U.S.A.

Paris, T.R. and B. Duff, 1988. A Methodology For Integrating Women's Concerns into Rice-based Farming Systems Research: An IRRI Case Study. Paper Presented at the International Conference on Appropriate Agricultural Technologies for Farm Women, Vigyan Bhawan, New Delhi, India. Sponsored by ICAR in collaboration with IRRI. Manila, Philippines.

Parthasarthy, P.B. and G. Satyanarayan, 1987. "Integrated Approach for Horticultural Development in Andhra Pradesh". Indian Journal of Agricultural Economics, Vol. XLII(3): 466-67.

Patel, A.S. and N.S. Jodha, 1979. "Economic Relationship Between Crop Farming and Dairying in Developing Area, A PL-480 Micro-level Study of South Gujarat". A.E.R.C., S.P. University, Deptt. of Economics, Gujarat, India.

Pathak, P.S. and R.D. Roy, 1987. "Silvipastural System Forage Production on Wastelands". Indian Farming, 37 (8): 10-11.

Patil, S.D., 1978. "An Economic Analysis of Farming System in a Hilly Tract of Western Maharashtra". Indian Journal of Agricultural Economics, Vol. XXXIII(4): pp. 59.

Planning Commission, 1989. "Agro-Climatic Regional Planning—An Overview". New Delhi: Government of India.

Planning Commission, 1990. "Approach to the Eighth Five Year Plan 1990-95". New Delhi: Government of India.

Radhakrishana, D., 1962. "An Application of Linear programming for Farm Planning: A Case Study in West Godawari District of Andhra pradesh". Indian Journal of Agricultural Economics. 17(2): 73-81.

Raynolds, S.G. and P. Lund, 1981. "The Development of Integrated Farming Systems for the Small Holder in Zanzibar". Proceedings of the conference on farming systems and farming systems research, held at the international conference centre, Arusha, Tanzania, April 14-16: 159-164.

Reddy, Y.V.R. and G.R. Korwar, 1985. "A Comparative Economic Evaluation of Agro-Forestry, Silvo-Agricultural and Silvo-Pastoral Systems in Drylands". Agricultural Situation in India. Vol. XL(8): 705-708.

Ruthenberg, H., 1976. "Farming Systems in the Tropics" (2nd ed.). Oxford: Clarendon Press.

Salam, M.A., D. Sreekumar and M.K. Mammen, 1990. "Mixed Farming on Homestead Agricultural—An Economical Approach". Indian Farming, 40(5): 14-16.

Sampath, T.V., M. Pal and I.P.S. Ahlawat, 1987. "New Approach to Cropping Systems". Indian Farming, 37(2): 16-19.

Savale, R.S., 1966. "Intensive Development Approach to Agricultural

Development: Role of Irrigation and Cropping Pattern in Agricultural Development". Indian Journal of Agricultural Economics, 21(4): 96-104.

Sexena, S.K., 1988. "Present Status of Common Grazing Lands of Indian Desert and their Possible Rangeland resource management". Proceedings of the National Rangeland Management Symposium, IGFRI, Jhansi Nov. 9-12 (Eds. Singh, P. and P.S. Pathak): 84-92.

Shah, S.L., 1979. "Farming Systems in Hill Areas". Summaries of the Group Discussions. Indian Journal of Agricultural Economics. Vol. XXXIV(I): 19-20.

Sharma, A.K., 1990. Economics of Different Irrigation Systems in Kangra District of Himachal Pradesh. Ph.D. Thesis. Department of Agricultural Economics, HPKV, Palampur: 276 pp.

Sharma, L.R. and J.P. Bhati, 1995. Role of Forests in Mountain Farming Systems of Himachal Pradesh. Paper Presented in International Seminar on Natural Resource Management and Their Linkages with Farming Systems, June 8-9, UHF, Nauni, Solan (H.P.).

Sharma, L.R., J.P. Bhati and R. Singh, 1991. "Emerging Farming Systems in Himachal Pradesh Himalaya: Key Issues in Sustainability". Indian Journal of Agricultural Economics. Vol. 46, No. 3, pp. 422-427.

Sharma L.R., R. Chand and J.P. Bhati, 1989. "An Analysis of Farmers Dependence on Forest for Fuelwood, Fodder and Timber in Himachal Pradesh". Agricultural Situation in India, 44(5): 359-362.

Singh, D., 1995. Changing Size and Status of Common Property Resources and Ecological Stress in Himachal Pradesh. Paper Presented in International Seminar on Natural Resources Management and their Linkages with Farming Systems, June 8-9, UHF. Nauni-Solan (H.P.).

Singh, D.V. and J.P. Bhati, 1995. Role Common Pool Resources in Household Energy in Himachal Pradesh. Paper Presented in International Seminar on Natural Resources Management and Their Likages with Farming Systems, June 8-9, UHF, Nauni-Solan (H.P.).

Singh, K. and B.S. Dhillon, 1995. Management of Natural Resources and its Impact on Socio-Economic Life of the Rural People in Punjab. Paper Presented in International Seminar on Natural Resource Management and Their Linkages with Farming Systems, June 8-9, UHF, Nauni-Solan (H.P.).

Singh, K. and V. Ballabh, 1989. "Afforestation of Village Common Lands: A case Study of Aslali Village Woodlot in Gujrat". Institute of Rural Management, Anand: 28pp.

Singh, R.P., B.L. Jain and V. Bhasker Rao, 1984. "Features of Traditional Farming Systems in Selected Villages of Madhya Pradesh". ICRISAT, Economics Progress Report-61, Patancheru, A.P., India.

Singh, R., R.K. Patel and S.S. Ahlawat, 1979. "Impact of Integrated Crop and Milk Production on Small Farms in Punjab". Indian Journal of Agricultural Economics. Vol. XXXII(3): 136-143.

Singh, R.P. and S.B. Singh, 1982. "Features of Traditional Farming Systems in Two Villages of Gujarat". ICRISAT, Economics Progress Report-35, Patancheru. A.P., India.

Singh, V. and V. Singh, 1989. "Improving the Productivity of Grass Land Based Farming Systems in Uttar Pradesh Hills". Advances in Forestry Research in India. Vol. III, International Book Distributors, Dehradun, p. 157-176.

Smith, M.M., 1986. "The Impact of Changing Agricultural Systems on the Nutritional Status of Farm Households in Developing Countries". Food and Nutrition Bulletin, 8(3): 25-29.

Speeding, C.R.W., 1979. An Introduction to Agricultural Systems. Applied Science Publishers Ltd., London.

Suryavanshi, S.D., 1979. "Economics of Farming Systems in Tribal Area of Northern Maharashtra". Indian Journal of Agriculture Economics, Vol. XXXIII(4): pp. 60.

Tewari, S.C., 1985. "Ameliorating Farm Income by Agro-Forestry on Marginal and Sub-Marginal Lands in the Lower Hills of H.P.". Agro-Forestry System: A Challenge. Edited by P.K. Khosla and Sunil Puri, ISTS: 185-190.

Thakur, D.R., K.D. Sharma and D.C. Thakur, 1995. Impact of Sheep and Goats Farming on Natural Resources of Himachal Pradesh. Paper presented in International Seminar on Natural Resources Management and Their Linkages with Farming Systems, June 8-9, UHF, Nauni-Solan (H.P.).

Tothill, J.C., 1988. "Fodder and Forage Management for Small Holder Farms in Euthopean Highlands". Euthopean Journal of Agricultural Science, 10(1-2): 71-81.

Vaidya, C.S. and R. Singh, 1995. Sustainability of Farm in Different Regions of Himachal Pradesh. Paper presented in International Seminar on Natural Resources Management and Their Linkages with Farming Systems held at Nauni (Solan) on June 8-9.

Wade, R., 1988. "Village Republics: Economic Conditions for Collective Action in South India". Cambridge South Asian Studies, Cambridge University Press: 238pp.

Walker, T.S., R.P. Singh and N.S. Jodha, 1983. "Dimensions of Farm-Level Diversification in the Semi-Arid Tropics of Rural South India". ICRISAT, Economics Progress Report-51, Patancheru, A.P., India.

Index